SOLUTIONS MANUAL TO ACCOMPANY FUNDAMENTALS OF CALCULUS

SOLUTIONS MANUAL TO ACCOMPANY FUNDAMENTALS OF CALCULUS

CARLA C. MORRIS
University of Delaware

ROBERT M. STARK
University of Delaware

Published by John Wiley & Sons, Inc., Hoboken, New Jersey
Published simultaneously in Canada

Library of Congress Cataloging-in-Publication Data:

Morris, Carla C.
Fundamentals of Calculus / Carla C. Morris, Robert M. Stark.
pages cm
Includes bibliographical references and index.
ISBN 978-1-119-01526-0 (cloth)
1. Calculus–Textbooks. I. Stark, Robert M., 1930- II. Title.
QA303.2.M67 2015
515–dc23

2014042182

10 9 8 7 6 5 4 3 2 1

CONTENTS

CHAPTER 1

LINEAR EQUATIONS AND FUNCTIONS

EXERCISES 1.1

1. $3x + 1 = 4x - 5$

$1 = x - 5$ conditional equation

$x = 6$

3. $5(x + 1) + 2(x - 1) = 7x + 6$

$5x + 5 + 2x - 2 = 7x + 6$

$7x + 3 = 7x + 6$ contradiction

5. $4(x + 3) = 2(2x + 5)$

$4x + 12 = 4x + 10$ contradiction

7. $5x - 3 = 17$

$5x = 20$

$x = 4$

Solutions Manual to Accompany Fundamentals of Calculus, First Edition. Carla C. Morris and Robert M. Stark.

9.
$$2x = 4x - 10$$
$$2x - 4x = -10$$
$$-2x = -10$$
$$x = 5$$

11.
$$4x - 5 = 6x - 7$$
$$-5 + 7 = 6x - 4x$$
$$2 = 2x$$
$$1 = x$$

13.
$$0.6x = 30$$
$$x = \frac{30}{0.60} = 50$$

15.
$$\frac{2}{3} = \left(\frac{4}{5}\right)x - \frac{1}{3}$$
$$15\left(\frac{2}{3}\right) = 15\left\{\left(\frac{4}{5}\right)x - \frac{1}{3}\right\}$$
$$10 = 12x - 5$$
$$15 = 12x$$
$$\frac{5}{4} = x$$

17.
$$5(x - 4) = 2x + 3(x - 7)$$
$$5x - 20 = 2x + 3x - 21$$
$$5x - 20 = 5x - 2$$
No solution

19.
$$3s - 4 = 2s + 6$$
$$s - 4 = 6$$
$$s = 10$$

21.
$$7t + 2 = 4t + 11$$
$$7t - 4t = 11 - 2$$
$$3t = 9$$
$$t = 3$$

23. $4(x+1)+2(x-3)=7(x-1)$

$$4x+4+2x-6=7x-7$$
$$6x-2=7x-7$$
$$6x-7x=-7+2$$
$$-x=-5$$
$$x=5$$

25. $\frac{x+8}{2x-5}=2$

$$(x+8)=2(2x-5)$$
$$x+8=4x-10$$
$$8+10=4x-x$$
$$18=3x$$
$$6=x$$

27. $8-\{4[x-(3x-4)-x]+4\}=3(x+2)$

$$8-\{4[x-3x+4-x]+4\}=3x+6$$
$$8-\{4[-3x+4]+4\}=3x+6$$
$$8-\{-12x+16+4\}=3x+6$$
$$8-\{-12x+20)=3x+6$$
$$8+12x-20=3x+6$$
$$12x-12=3x+6$$
$$9x=18$$
$$x=2$$

29. $6x-3y=9$ for x

$$6x=3y+9$$
$$x=\frac{3y+9}{6}=\frac{1}{2}y+\frac{3}{2}$$

31. $3x+5y=15$

$$5y=15-3x$$

$\longrightarrow$

$$y = \frac{(15 - 3x)}{5}$$

$$y = 3 - \left(\frac{3}{5}\right)x$$

33. $V = LWH$

$$\frac{V}{LH} = W$$

35. $Z = \dfrac{(x - \mu)}{\sigma}$

$$Z\sigma = x - \mu$$

$$x = Z\sigma + \mu$$

37. Let $x =$ monthly installment ($). Since Sally paid $300, she owes $1300 – $300 = $1000. Therefore, $5x = 1000$ or $x =$ $200 monthly installment.

39. The consumption function is $C(x) = mx + b$. The slope is the “marginal propensity to consume.” Therefore, $C(x) = 0.75x + b$. The disposable income $x = 2$ for a consumption $y = 11$ yields $11 = (0.75)(2) + b$, so $b = 9.5$ and consumption is $C(x) = 0.75x + 9.5$.

41. a) $d = 4.5(2) = 9$ miles

b) $18 = 4.5t$ and $t = 18/4.5 = 4$ seconds

43. The tax is 6.2%, or 0.062 as a decimal form, so $T = 0.062x$, where $0 \leq x \leq 87{,}000$.

45. a) $\text{BSA} = 1321 + (0.3433)(20{,}000) = 8187\,\text{cm}^2$

b) $1330 = 1321 + (0.3433)(\text{Wt})$

$9 = (0.3433)(\text{Wt})$

$9/0.3433 = 26.2\ \text{kg} = \text{Wt}.$

EXERCISES 1.2

1. Setting $y = 0$ determines the x-intercept and setting $x = 0$ determines the y-intercept.

a) $5x - 3y = 15$ x-intercept 3, y-intercept -5

b) $y = 4x - 5$ x-intercept 5/4, y-intercept -5

c) $2x + 3y = 24$ x-intercept 12, y-intercept 8

⟶

d) $9x - y = 18$ x-intercept 2, y-intercept -18

e) $x = 4$ x-intercept 4, no y-intercept(vertical line)

f) $y = -2$ no x-intercept (horizontal line), y-intercept -2

3. The slope is $m = \dfrac{y_2 - y_1}{x_2 - x_1}$

a) (3, 6) and (−1, 4) $m = \dfrac{4-6}{-1-3} = \dfrac{-2}{-4} = \dfrac{1}{2}$

b) (1, 6) and (2, 11) $m = \dfrac{11-6}{2-1} = \dfrac{5}{1} = 5$

c) (6, 3) and (12, 7) $m = \dfrac{7-3}{12-6} = \dfrac{4}{6} = \dfrac{2}{3}$

d) (2, 3) and (2, 7) $m = \dfrac{7-3}{2-2} = \dfrac{4}{0}$ undefined

e) (2, 6) and (5, 6) $m = \dfrac{6-6}{5-2} = \dfrac{0}{3} = 0$

f) (5/3, 2/3) and (10/3, 1) $m = \dfrac{1 - \frac{2}{3}}{\frac{10}{3} - \frac{5}{3}} = \dfrac{\frac{1}{3}}{\frac{5}{3}} = \dfrac{1}{5}$

5. a) x-intercept 5/2 and y-intercept -5

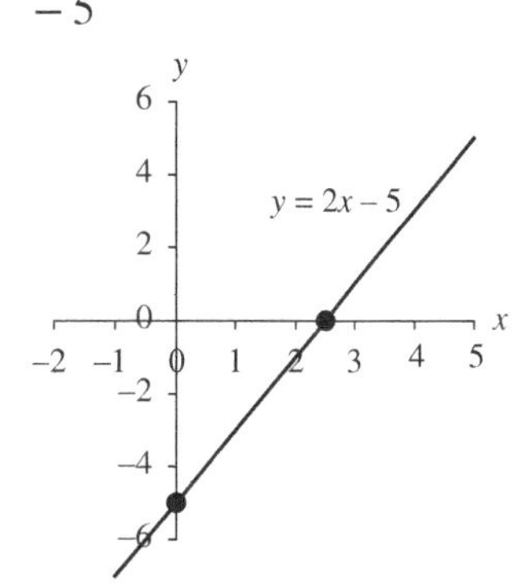

b) x-intercept 4 and no y-intercept

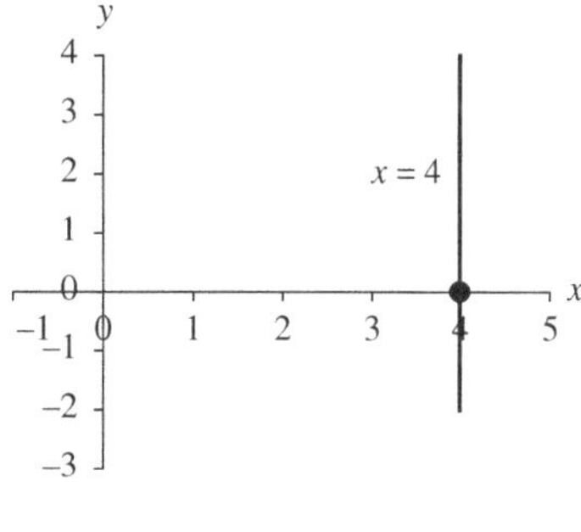

c) x-intercept 5 and y-intercept 3

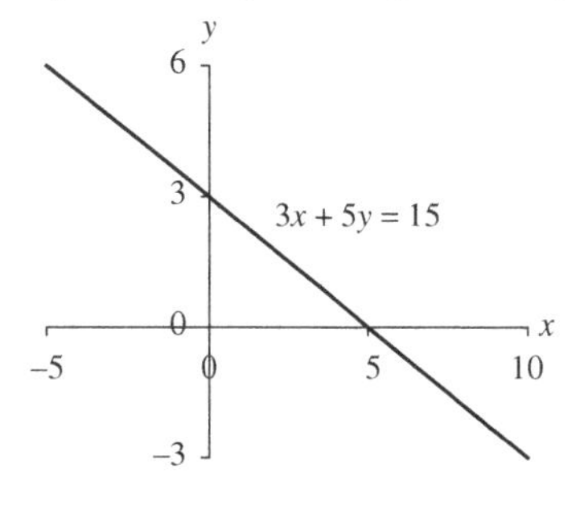

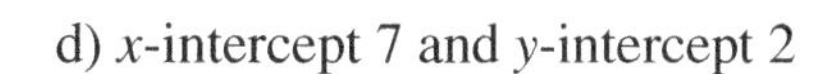
d) x-intercept 7 and y-intercept 2

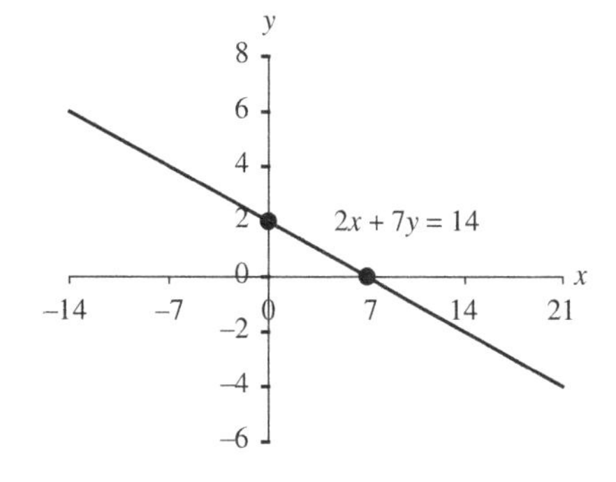

7. a) For $y = (5/3)x + 2$ and $5x - 3y = 10$; the slope of the first line is 5/3. Solving for y in the second equation yields $y = (5/3)x - (10/3)$. This slope is also 5/3.
The lines are parallel (same slope, different intercepts).

b) $6x + 2y = 4$ and $y = (1/3)x + 1$. The slope of the second line is apparent from the slope intercept form as 1/3. Solve for y in the first equation as $y = -3x + 2$. The slope is -3. The slopes are negative reciprocals, so the lines are perpendicular.

c) $2x - 3y = 6$ and $4x - 6y = 15$. Solving for y in each equation, yields $y = (2/3)x - 2$ and $y = (2/3)x - (5/2)$. These lines have the same slope (and different intercepts) so they are parallel.

d) $y = 5x - 4$ and $3x - y = 4$. The slope of the first line is 5. In the second equation, ($y = 3x - 4$), the slope is 3. These slopes are neither the same nor negative reciprocals. The lines are neither parallel nor perpendicular.

e) $y = 5$ is a horizontal line while $x = 3$ is a vertical line.
The two lines are perpendicular.

9. Generally, lines have a single x-intercept. The exception $y = 0$ (the x-axis) with an infinite number of x-intercepts. Any horizontal line (except $y = 0$) has no x-intercepts. Generally, lines do not have more than one y-intercept. The exception $x = 0$ (the y-axis) with an infinite number of y-intercepts. Any vertical line (except $x = 0$) has no y-intercepts.

11. The ordered pairs with time and machine values are (0, 75,000) and (9, 21,000). The slope $m = \dfrac{21{,}000 - 75{,}000}{9 - 0} = \dfrac{-54{,}000}{9} = -6000.$
The y-intercept is the initial cost, \$75,000. Therefore, to model the straight-line depreciation $V(t) = -6000t + 75{,}000$ where $V(t)$ is the machine's value (\$) at time t.

13. Ordered pairs (gallons of gasoline, miles traveled) are (7, 245) and (12, 420). The slope is $\dfrac{420 - 245}{12 - 7} = \dfrac{175}{5} = 35$. Let $x =$ gallons of gasoline and $y =$ miles traveled. Then either ordered pair, with the point slope formula, yields $y - 245 = 35(x - 7)$ or $y = 35x$.

15. Total cost is fixed plus variable costs. The fixed cost is monthly rent of \$1100. The variable cost is $5x$, where x is monthly production.

⟶

The total cost is $C(x) = 1100 + 5x$.

17. a) Here, fixed cost is ($50/day) and variable cost is ($0.28/mile). So, $C(x) = 50 + 0.28x$

b) If one has $92, the equation for the travel distance is $92 = 50 + 0.28x$

Solving,

$$42 = 0.28x$$
$$\frac{42}{0.28} = x$$
$$150 = x$$

The car can be rented and driven 150 miles for $92.

19. Since R is a function of C, the ordered pairs (C, R) are $(70, 84)$ and $(40, 48)$. The slope is $\frac{48-84}{40-70} = \frac{36}{30} = \frac{6}{5}$. Either ordered pair determines the equation as the slope is known.

Therefore, $R - 84 = (6/5)(C - 70)$ or, $R = (6/5)C$.

EXERCISES 1.3

1. Here, the GCF is 8, so $8x - 24 = 8(x - 3)$

3. Here, the GCF is $5x$, so $5x^3 - 10x^2 + 15x = 5x(x^2 - 2x + 3)$

5. Here, the GCF is $5a^3bc^3$, so $5a^3b^2c^4 + 10a^3bc^3 = 5a^3bc^3(bc + 2)$

7. Here, the GCF is $5x^2y^3z^5$, so $20x^3y^5z^6 + 15x^4y^3z^7 + 20x^2y^4z^5 = 5x^2y^3z^5(4xy^2z + 3x^2z^2 + 4y)$

9. This is a difference of squares, so $x^2 - 25 = (x - 5)(x + 5)$

11. There is a GCF of 3 to yield $3(x^2 + 9)$. A sum of squares is not factorable.

13. There is a GCF of 2 to yield $2(x^3 - 8)$. Next, using the difference of cubes formula the expression factors as $2(x - 2)(x^2 + 2x + 4)$.

15. There is a GCF of $7(a + b)$ to yield $7(a + b)(x^2 - 4)$. Next, using the difference of squares formula yields $7(a + b)(x + 2)(x - 2)$.

17. The last term is +4 and since the middle term is positive, one seeks two positive factors of 4 that add to 5. The expression factors as $x^2 + 5x + 4 = (x + 4)(x + 1)$.

19. The last term is positive, and since the middle term is positive, one seeks two positive factors of 1 that add to 3. This is not possible. Therefore, $x^2 + 3x + 1$ is not factorable.

21. Here, the last term is negative so seek one positive factor and one negative factor of 16 that add to give −6. Therefore, $x^2 - 6x - 16 = (x - 8)(x + 2)$.

23. First, the GCF is 2 so $2x^2 + 12x + 16 = 2(x^2 + 6x + 8)$. Next, two positive factors of 8 that add to 6 are needed. The expression is completely factored as

$$2x^2 + 12x + 16 = 2(x + 4)(x + 2).$$

25. Seek two positive factors of 20 that add to 9. The expression factors as

$$a^2b^2 + 9ab + 20 = (ab + 4)(ab + 5).$$

27. First, the GCF is 2 so $2x^2y^2 + 28xy + 90 = 2(x^2y^2 + 14xy + 45)$. Next, seek two positive factors of 45 that add to 14. The expression factors as $2(xy + 9)(xy + 5)$.

29. Seek two positive factors of 5 that add to 7. Since this is not possible, the expression $x^2 + 7x + 5$ is prime.

31. This is a quadratic in x^2. Seek two negative factors of 4 that add to 5. Therefore, $x^4 - 5x^2 + 4 = (x^2 - 4)(x^2 - 1)$. The factors are both differences of squares, so factoring yields $x^4 - 5x^2 + 4 = (x - 2)(x + 2)(x - 1)(x + 1)$.

33. First, group the expression by powers to yield $(x^2 - a^2) + (5x - 5a)$. Then, factor each pair to yield $(x - a)(x + a) + 5(x - a)$. Next, a GCF of $(x - a)$ is factored from the expression to yield $(x - a)[(x + a) + 5]$.

35. First, the GCF is 2, so $4ab - 8ax + 6by - 12xy = 2[2ab - 4ax + 3by - 6xy]$. Group by pairs as $2[(2ab - 4ax) + (3by - 6xy)]$. Factoring each pair yields $2[2a(b - 2x) + 3y(b - 2x)] = 2[(b - 2x)(2a + 3y)]$.

37. Using $a = 1$, $b = 9$, and $c = 8$ in the quadratic formula yields $x = \dfrac{-(9) \pm \sqrt{(9)^2 - 4(1)(8)}}{2(1)} = \dfrac{-9 \pm \sqrt{49}}{2} = \dfrac{-9 \pm 7}{2}$. So $x = -8$ or $x = -1$.

39. Using $a = 1$, $b = 17$, and $c = 72$ in the quadratic formula yields $x = \dfrac{-(17) \pm \sqrt{(17)^2 - 4(1)(72)}}{2(1)} = \dfrac{-17 \pm \sqrt{1}}{2} = \dfrac{-17 \pm 1}{2}$. So, $x = -8$ or $x = -9$.

41. Using $a = 1$, $b = 4$, and $c = 7$ in the quadratic formula yields $x = \dfrac{-(4) \pm \sqrt{(4)^2 - 4(1)(7)}}{2(1)} = \dfrac{-4 \pm \sqrt{-12}}{2}$. There are no real solutions.

43. First, rewrite as $x^2 - 9x + 18 = 0$. Next, using $a = 1$, $b = -9$, and $c = 18$ in the quadratic formula yields $x = \dfrac{-(-9) \pm \sqrt{(-9)^2 - 4(1)(18)}}{2(1)} = \dfrac{9 \pm \sqrt{9}}{2} = \dfrac{9 \pm 3}{2}$. So, $x = 6$ or $x = 3$.

45. Using $a = 2$, $b = -3$, and $c = 1$ in the quadratic formula yields $x = \dfrac{-(-3) \pm \sqrt{(-3)^2 - 4(2)(1)}}{2(2)} = \dfrac{3 \pm \sqrt{1}}{4}$. So, $x = 1$ or $x = 1/2$.

EXERCISES 1.4

1. (3, 7)

3 7

3. (5, ∞)

5

5. (−2, 1)

−2 1

7. $(4, \infty)$

9. $(-3, 7)$

11. $[1, 8)$

13. $[5, 8)$

15. All real numbers

17. $[5/2, \infty)$

19. $(-\infty, -3) \cup (-3, 1) \cup (1, \infty)$

21. a) $f(0) = 3$

b) $f(1) = 15$

c) $f(x + 3) = 7(x + 3)^3 + 5(x + 3) + 3$

23. a) $f(-1) = 10$

b) $f(a^2) = a^{10} + 11$

c) $f(x + h) = (x + h)^5 + 11$

25. It is not a function. It fails the vertical line test.

27. It is not a function. It fails the vertical line test.

29. $f(x) = x^2 - 4$

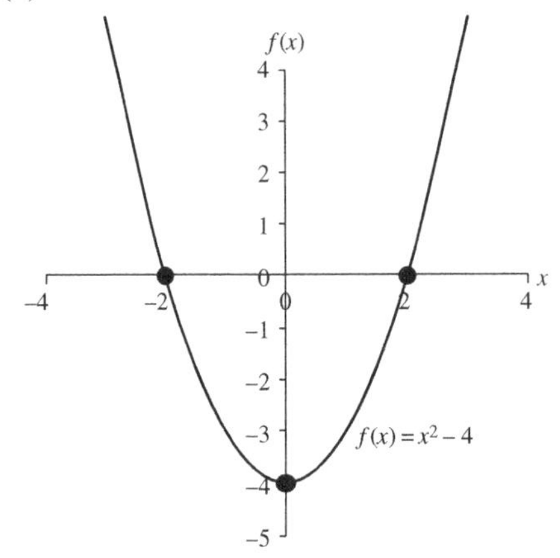

31. $f(x) = x^3 - 8$

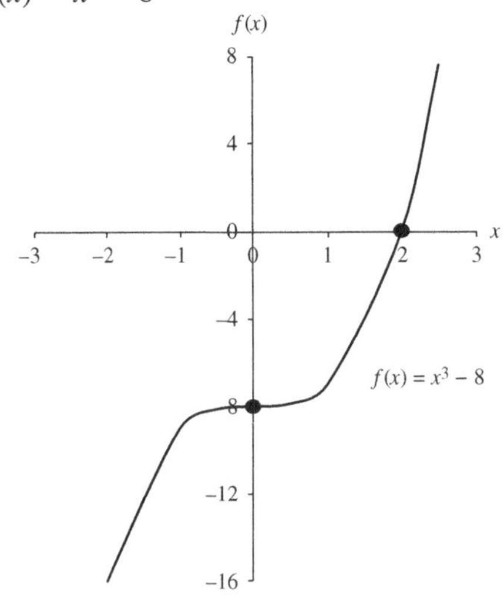

33. It is the piecewise graph.

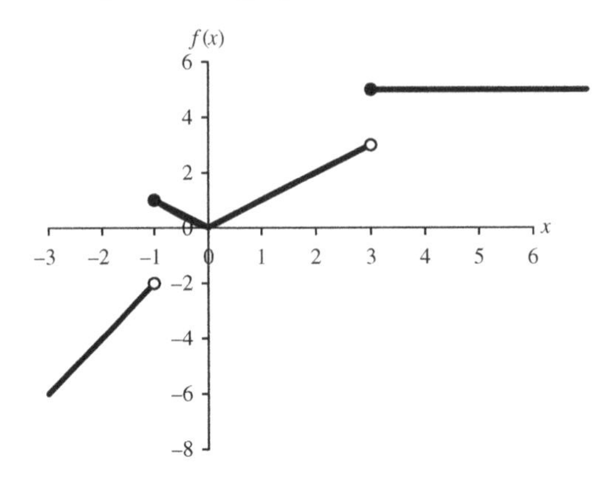

35. a) $(3x^5 + 7x^3 + 8) - (4x^5 - 2x^3 + 2x) = -x^5 + 9x^3 - 2x + 8$

b) $(4x^5 - 2x^3 + 2x) - (3x^5 + 7x^3 + 8) = x^5 - 9x^3 + 2x - 8$

c) $(4x^5 - 2x^3 + 2x)(3x^5 + 7x^3 + 8)$

d) $3(4x^5 - 2x^3 + 2x)^5 + 7(4x^5 - 2x^3 + 2x)^3 + 8$

37. a) $2x^5 + h$

b) $(x + h)^2 + 4$

c) $(2a^5)(a^2 + 4)$

d) $2(x + 1)^5[(x + 2)^2 + 4]$

EXERCISES 1.5

1. $1^{3/7} = \left(\sqrt[7]{1}\right)^3 = 1$

3. $(25)^{3/2} = (\sqrt{25})^3 = (5)^3 = 125$

5. $(64)^{5/6} = \left(\sqrt[6]{64}\right)^5 = (2)^5 = 32$

7. $\left(\frac{2}{3}\right)^{-2} = \left(\frac{3}{2}\right)^2 = \frac{9}{4}$

9. $(0.008)^{1/3} = \sqrt[3]{0.008} = 0.20$

11. $\frac{15^3}{5^3} = \left(\frac{15}{5}\right)^3 = 3^3 = 27$

13. $x^3x^5 = x^8$

15. $(2xy)^3 = 2^3x^3y^3 = 8x^3y^3$

17. $\frac{x^3x^5}{x^{-4}} = \frac{x^8}{x^{-4}} = x^{12}$

19. $\frac{x^4y^5}{x^2y^{-2}} = x^{4-2}y^{5-(-2)} = x^2y^7$

21. $\left(\frac{2x^3}{y^2}\right)^2 = \frac{(2)^2(x^3)^2}{(y^2)^2} = \frac{4x^6}{y^4}$

23. $\sqrt[3]{x^5}\sqrt[3]{x^4} = \sqrt[3]{x^9} = x^3$

25. $(81x^4y^8)^{1/4} = \sqrt[4]{81x^4y^8} = 3xy^2$

27. $\dfrac{(16x^4y^5)^{3/2}}{\sqrt{y}} = \dfrac{(16)^{3/2}x^6y^{15/2}}{y^{1/2}} = (\sqrt{16})^3x^6y^7 = 64x^6y^7$

29. $\dfrac{(8x^5y^7)^{2/3}}{\sqrt[3]{xy^2}} = \dfrac{(8)^{2/3}x^{10/3}y^{14/3}}{x^{1/3}y^{2/3}} = 4x^3y^4$

EXERCISES 1.6

1. $f(x) = 6x + 11$

a) $\dfrac{f(x+h) - f(x)}{h} = \dfrac{[6(x+h) + 11] - [6x + 11]}{h} = \dfrac{6h}{h} = 6$

b) $\dfrac{f(x) - f(a)}{x - a} = \dfrac{[6x + 11] - [6a + 11]}{x - a} = \dfrac{6x - 6a}{x - a} = \dfrac{6(x - a)}{x - a} = 6$

3. $f(x) = 7x - 4$

a) $\dfrac{f(x+h) - f(x)}{h} = \dfrac{[7(x+h) - 4] - [7x - 4]}{h} = \dfrac{7h}{h} = 7$

b) $\dfrac{f(x) - f(a)}{x - a} = \dfrac{[7x - 4] - [7a - 4]}{x - a} = \dfrac{7x - 7a}{x - a} = \dfrac{7(x - a)}{x - a} = 7$

5. $f(x) = x^2 - 7x + 4$

a) $\dfrac{f(x+h) - f(x)}{h} = \dfrac{[(x+h)^2 - 7(x+h) + 4] - [x^2 - 7x + 4]}{h}$

$\dfrac{2xh + h^2 - 7h}{h} = \dfrac{h(2x + h - 7)}{h} = 2x + h - 7$

b) $\dfrac{f(x) - f(a)}{x - a} = \dfrac{[x^2 - 7x + 4] - [a^2 - 7a + 4]}{x - a}$

$= \dfrac{(x^2 - a^2) - (7x - 7a)}{x - a}$

$= \dfrac{(x - a)(x + a) - 7(x - a)}{x - a} = x + a - 7$

7. $f(x) = x^2 + 6x - 8$

a) $\dfrac{f(x+h) - f(x)}{h} = \dfrac{[(x+h)^2 + 6(x+h) - 8] - [x^2 + 6x - 8]}{h}$

$\dfrac{2xh + h^2 + 6h}{h} = \dfrac{h(2x + h + 6)}{h} = 2x + h + 6$

b) $\dfrac{f(x) - f(a)}{x - a} = \dfrac{[x^2 + 6x - 8] - [a^2 + 6a - 8]}{x - a}$

$= \dfrac{(x^2 - a^2) + (6x - 6a)}{x - a}$

$= \dfrac{(x - a)(x + a) + 6(x - a)}{x - a} = x + a + 6$

9. $f(x) = 5x^2 - 2x - 3$

a) $\dfrac{f(x+h)-f(x)}{h} = \dfrac{[5(x+h)^2 - 2(x+h) - 3] - [5x^2 - 2x - 3]}{h}$

$\dfrac{10xh + 5h^2 - 2h}{h} = \dfrac{h(10x + 5h - 2)}{h} = 10x + 5h - 2$

b) $\dfrac{f(x)-f(a)}{x-a} = \dfrac{[5x^2 - 2x - 3] - [5a^2 - 2a - 3]}{x-a}$

$= \dfrac{(5x^2 - 5a^2) - (2x - 2a)}{x-a}$

$= \dfrac{5(x-a)(x+a) - 2(x-a)}{x-a} = 5x + 5a - 2$

11. $f(x) = x^3 - 4x + 5$

a) $\dfrac{f(x+h)-f(x)}{h} = \dfrac{[(x+h)^3 - 4(x+h) + 5] - [x^3 - 4x + 5]}{h}$

$\dfrac{3x^2h + 3xh^2 + h^3 - 4h}{h} = \dfrac{h(3x^2 + 3xh + h^2 - 4)}{h}$

$= 3x^2 + 3xh + h^2 - 4$

b) $\dfrac{f(x)-f(a)}{x-a} = \dfrac{[x^3 - 4x + 5] - [a^3 - 4a + 5]}{x-a}$

$= \dfrac{(x^3 - a^3) - (4x - 4a)}{x-a}$

$= \dfrac{(x-a)(x^2 + ax + a^2) - 4(x-a)}{x-a}$

$= x^2 + ax + a^2 - 4$

13. $f(x) = 2x^3 - 7x + 3$

a) $\dfrac{f(x+h)-f(x)}{h} = \dfrac{[2(x+h)^3 - 7(x+h) + 3] - [2x^3 - 7x + 3]}{h}$

$\dfrac{6x^2h + 6xh^2 + 2h^3 - 7h}{h} = \dfrac{h(6x^2 + 6xh + 2h^2 - 7)}{h}$

$= 6x^2 + 6xh + 2h^2 - 7$

b) $\dfrac{f(x)-f(a)}{x-a} = \dfrac{[2x^3 - 7x + 3] - [2a^3 - 7a + 3]}{x-a}$

$= \dfrac{(2x^3 - 2a^3) - (7x - 7a)}{x-a}$

$= \dfrac{2(x-a)(x^2 + ax + a^2) - 7(x-a)}{x-a}$

$= 2(x^2 + ax + a^2) - 7$

15. $f(x) = \dfrac{3}{x^3}$

a)

$$\frac{f(x+h)-f(x)}{h} = \frac{\dfrac{3}{(x+h)^3} - \dfrac{3}{x^3}}{h} = \frac{3x^3 - 3(x+h)^3}{x^3h(x+h)^3}$$

$$= \frac{-9x^2h - 9xh^2 - 3h^3}{x^3h(x+h)^3}$$

$$= \frac{-3h(3x^2+3xh+h^2)}{x^3h(x+h)^3} = \frac{-3(3x^2+3xh+h^2)}{x^3(x+h)^3}$$

b)

$$\frac{f(x)-f(a)}{x-a} = \frac{\dfrac{3}{x^3} - \dfrac{3}{a^3}}{x-a} = \frac{\dfrac{3a^3-3x^3}{a^3x^3}}{x-a} = \frac{3(a-x)(a^2+ax+x^2)}{(x-a)a^3x^3}$$

$$= \frac{-3(a^2+ax+x^2)}{a^3x^3}$$

SUPPLEMENTARY EXERCISES CHAPTER 1

1. $9(x-3) + 2x = 3(x+1) - 2$

$$9x - 27 + 2x = 3x + 3 - 2$$

$$11x - 27 = 3x + 1$$

$$8x = 28$$

$$x = \frac{7}{2}$$

3. $Z = \dfrac{x-\mu}{\sigma}$ so $Z\sigma = x - \mu$ and $\mu = x - Z\sigma$.

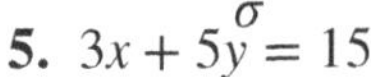

5. $3x + 5y = 15$

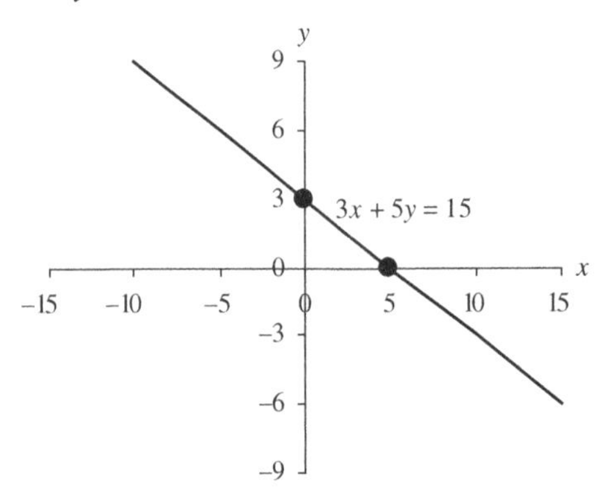

7. First, find the slope $m = \frac{1-7}{2-5} = \frac{-6}{-3} = 2$. Next, use the point slope form with either point to yield $y - 7 = 2(x - 5)$.

9. Rewriting in slope intercept form $4x - 3y = 12$ as $y = \frac{4}{3}x - 4$. The slope is $\frac{4}{3}$. A line parallel has the same slope, so the line of interest is $y - 5 = \frac{4}{3}(x - 2)$.

11. The GCF of $2x^3 - 18x^2 - 20x$ is $2x$, so the initial factoring yields $2x(x^2 - 9x - 10)$. Complete factoring yields $2x(x - 10)(x + 1)$.

13. Group as $(2ax - 2ay) + (bx - by)$ before factoring the GCF from each pair of terms. Then $2a(x - y) + b(x - y) = (x - y)(2a + b)$ is the completely factored expression.

15. Multiplying by 3 to eliminate the fraction yields $2(x - 1) < 3(x - 2)$. Next, $2x - 2 < 3x - 6$ yields $4 < x$ so the interval notation solution is $(4, \infty)$.

17. To determine the domain, factor the denominator. This yields $\frac{2x+5}{x(x+8)(x+1)}$. The domain is all real numbers except $x = 0$, $x = -8$, or $x = -1$.

19. a) $(f - g)(x) = -x^3 + 4x^2 + 3x + 5$

b) $(f \cdot g)(x) = (x^2 + 3x + 1)(x^3 - 3x^2 - 4)$

c) $(f \circ g)(x) = f(g(x)) = (x^3 - 3x^2 - 4)^2 + 3(x^3 - 3x^2 - 4) + 1$

21. $\left(\frac{-2x^5}{3x^2}\right) \cdot \left(\frac{3x^{-2}}{4x^5}\right) = \frac{-6x^3}{12x^7} = \frac{-1}{2x^4}$

23. $\left(\frac{2x^3}{3y^{-2}z^4}\right)^{-3} = \left(\frac{2x^3y^2}{3z^4}\right)^{-3} = \left(\frac{3z^4}{2x^3y^2}\right)^3 = \frac{27z^{12}}{8x^9y^6}$

25. $f(x) = x^3 + 3x + 1$

a) $$\frac{f(x+h) - f(x)}{h} = \frac{[(x+h)^3 + 3(x+h) + 1] - [x^3 + 3x + 1]}{h}$$
$$\frac{3x^2h + 3xh^2 + h^3 + 3h}{h} = \frac{h(3x^2 + 3xh + h^2 + 3)}{h}$$
$$= 3x^2 + 3xh + h^2 + 3$$

b) $$\frac{f(x) - f(a)}{x - a} = \frac{[x^3 + 3x + 1] - [a^3 + 3a + 1]}{x - a}$$
$$= \frac{(x^3 - a^3) + (3x - 3a)}{x - a}$$
$$= \frac{(x - a)(x^2 + ax + a^2) + 3(x - a)}{x - a} = x^2 + ax + a^2 + 3$$

CHAPTER 2

THE DERIVATIVE

EXERCISES 2.1

1. The slope of the tangent is the derivative, so $y' = 0$.
3. The derivative is $2x$. When $x = 5$, the derivative (slope) is 10.
5. The derivative is $2x$. When $x = 1/2$, the derivative (slope) is 1.
7. At $x = 5/2$, the derivative or slope is $2(5/2) = 5$.
The point is $(5/2, (5/2)^2)$ or (5/2, 25/4). Using point slope yields
$y - 25/4 = 5(x - 5/2)$.
9. At $x = -7/2$, the derivative (slope) is $2(-7/2) = -7$.
The point is $(-7/2, (7/2)^2)$ or $(-7/2, 49/4)$. Using point slope yields
$y - 49/4 = -7(x + 7/2)$.
11. One seeks the point where the derivative $2x = 7/4$.
Therefore, $2x = 7/4$ and $x = 7/8$. The point is (7/8, 49/64).
13. One seeks the point where the derivative $2x = 4/5$.
Therefore, $2x = 4/5$ and $x = 2/5$. The point is (2/5, 4/25).
15. The line has a slope of 2/3, so one seeks where the derivative is also 2/3. Therefore, $2x = 2/3$ so $x = 1/3$. The point is (1/3, 1/9).
17. Evaluating $3x^2$ at $x = 2$ yields a slope of 12.
19. Evaluating $3x^2$ at $x = 1/2$ yields a slope of 3/4.

Solutions Manual to Accompany Fundamentals of Calculus, First Edition. Carla C. Morris and Robert M. Stark.

21. Evaluating $3x^2$ at $x = 4$ yields a slope of 48. The point is $(4, (4)^3)$ or $(4, 64)$. The equation of the tangent line is $y - 64 = 48(x - 4)$.

23. Evaluating $3x^2$ at $x = 3/4$ yields a slope of 27/16. The point is $(3/4, (3/4)^3)$ or $(3/4, 27/64)$. The equation of the tangent line is $y - 27/64 = 27/16(x - 3/4)$.

25. One seeks where $3x^2 = 48$ or $x^2 = 16$. The two solutions are $x = 4$ and $x = -4$. The points are $(4, 64)$ and $(-4, -64)$.

27. Solving for y, the slope yields $y = (4/3)x - 2$ and a line parallel also has slope 4/3. Therefore, set $3x^2 = 4/3$. This yields $x = \pm\frac{2}{3}$ and the points $\left(\frac{2}{3}, \frac{8}{27}\right)$ and $\left(-\frac{2}{3}, -\frac{8}{27}\right)$.

EXERCISES 2.2

1. a) $\lim_{x\to 3-} f(x) = 10$ b) $\lim_{x\to 3+} f(x) = 4$ c) $\lim_{x\to 3} f(x)$ does not exist

3. a) $\lim_{x\to -2-} f(x) = -3$ b) $\lim_{x\to -2+} f(x) = -3$ c) $\lim_{x\to -2} f(x) = -3$

5. a) $\lim_{x\to 0-} f(x) = 0$ b) $\lim_{x\to 0+} f(x) = 0$ c) $\lim_{x\to 0} f(x) = 0$

7. $\lim_{x\to -2} 4 = 4$

9. $\lim_{x\to 5} (3x + 2) = 3(5) + 2 = 17$

11. $\lim_{x\to 4} (9x + 5) = 9(4) + 5 = 41$

13. $\lim_{x\to 1} (5x^2 + 9x + 3) = 5(1)^2 + 9(1) + 3 = 17$

15. $\lim_{x\to 0} (2x^7 + 3x^4 + 9x + 3) = 2(0)^7 + 3(0)^4 + 9(0) + 3 = 3$

17. $\lim_{x\to -1} (2x^3 + 5x^2 - 4) = 2(-1)^3 + 5(-1)^2 - 4 = -1$

19. $\lim_{x\to 2} \frac{3x}{x - 2} = \frac{3(2)}{2 - 2} = \frac{6}{0}$ is undefined, so the limit does not exist.

21. $\lim_{x\to 3} \frac{x^2 - 9}{x - 3} = \frac{3^2 - 9}{3 - 3} = \frac{0}{0}$. This is an indeterminate form. Try factoring as $\lim_{x\to 3} \frac{(x - 3)(x + 3)}{x - 3} = \lim_{x\to 3}(x + 3) = 3 + 3 = 6$.

23. $\lim_{x\to 2} \frac{x^3-8}{x-2} = \frac{2^3-8}{2-2} = \frac{0}{0}$. This is an indeterminate form.
Try factoring as

$$\lim_{x\to 2} \frac{(x-2)(x^2+2x+4)}{x-2} = \lim_{x\to 2}(x^2+2x+4) = 12$$

25. $\lim_{x\to\infty} \frac{-4}{x^3} = \frac{-4}{\infty^3}$ so the limit approaches 0.

27. $\lim_{x\to\infty}(2x^3+5x+1)$ approaches infinity. The limit does not exist.

29. $\lim_{x\to -1} f(x) = \frac{1}{-1} = -1$

31. $\lim_{x\to 1} f(x) = 2(1) + 5 = 7$

33. $\lim_{x\to 4} f(x) = (4)^2 - 3(4) - 4 = 0$

35. $\lim_{x\to 10} f(x) = 10$

EXERCISES 2.3

1. If $f(x) = 4x + 11$

a) $\lim_{h\to 0} \frac{[4(x+h)+11]-[4x+11]}{h} = \lim_{h\to 0} \frac{4h}{h} = 4 = f'(x).$

b) $\lim_{x\to a} \frac{[4x+11]-[4a+11]}{x-a} = \lim_{x\to a} \frac{4x-4a}{x-a}$

$= \lim_{x\to a} \frac{4(x-a)}{x-a} = 4 = f'(a).$

Therefore, $f'(x) = 4$.

3. If $f(x) = x^2 + 5x + 1$

a) $\lim_{h\to 0} \frac{[(x+h)^2+5(x+h)+1]-[x^2+5x+1]}{h}$

$= \lim_{h\to 0} \frac{2xh+h^2+5h}{h} = \lim_{h\to 0} \frac{h(2x+h+5)}{h} = 2x+5 = f'(x)$

b) $\lim_{x\to a} \frac{[x^2+5x+1]-[a^2+5a+1]}{x-a} = \lim_{x\to a} \frac{(x^2-a^2)+(5x-5a)}{x-a}$

$= \lim_{x\to a} \frac{(x-a)(x+a)+5(x-a)}{x-a} = 2a+5 = f'(a).$

Therefore, $f'(x) = 2x + 5$.

5. If $f(x) = x^2 - 6x + 1$

a) $$\lim_{h\to 0} \frac{[(x+h)^2 - 6(x+h) + 1] - [x^2 - 6x + 1]}{h}$$

$$= \lim_{h\to 0} \frac{2xh + h^2 - 6h}{h} = \lim_{h\to 0} \frac{h(2x + h - 6)}{h} = 2x - 6 = f'(x)$$

b) $$\lim_{x\to a} \frac{[x^2 - 6x + 1] - [a^2 - 6a + 1]}{x - a} = \lim_{x\to a} \frac{(x^2 - a^2) - (6x - 6a)}{x - a}$$

$$= \lim_{x\to a} \frac{(x-a)(x+a) - 6(x-a)}{x - a} = 2a - 6 = f'(a).$$

Therefore, $f'(x) = 2x - 6$.

7. If $f(x) = x^3 + 5$

a) $$\lim_{h\to 0} \frac{[(x+h)^3 + 5] - [x^3 + 5]}{h} = \lim_{h\to 0} \frac{3x^2h + 3xh^2 + h^3}{h}$$

$$= \lim_{h\to 0} \frac{h(3x^2 + 3xh + h^2)}{h} = 3x^2 = f'(x)$$

b) $$\lim_{x\to a} \frac{[x^3 + 5] - [a^3 + 5]}{x - a} = \lim_{x\to a} \frac{(x^3 - a^3)}{x - a}$$

$$= \lim_{x\to a} \frac{(x-a)(x^2 + ax + a^2)}{x - a} = 3a^2 = f'(a)$$

Therefore, $f'(x) = 3x^2$.

9. $f'(x) = 0$

11. $f'(x) = 0$

13. $f'(x) = 7$

15. $f'(x) = 14$

17. $f'(x) = 4x + 7$

19. $f'(x) = 30x^2 - 18x + 3$

21. $\frac{d}{dx}(7x^5 - 4x^4 + 3x^2 + 40) = 35x^4 - 16x^3 + 6x$

23. $\frac{d}{dx}\left(9x + \frac{4}{x}\right) = 9 - \frac{4}{x^2}$

25. $\frac{d}{dx}\left(\sqrt[6]{x^5} + \frac{2}{x^4} + 8\right) = \frac{5}{6}x^{-1/6} - \frac{8}{x^5}$

27. $y'(1) = 3(1)^2 + 4(1) + 3 = 10$

29. $f(2) = (2)^2 + 4(2) + 2 = 14$ and $f'(2) = 2(2) + 4 = 8$.

31. $f(-1) = (-1)^3 + 5(-1)^2 + 2 = 6$ and $f'(-1) = 3(-1)^2 + 10(1) = -7$.

33. $y'(2) = 6(2) + 5 = 17$ and $y(2) = 24$. The derivative is the slope of the tangent, so the equation is $y - 24 = 17(x - 2)$.

35. $y'(0) = 35(0)^4 + 8 = 8$ and $y(0) = 7(0)^5 + 8(0) + 25 = 25$.
The derivative is the slope of the tangent, so $y - 25 = 8(x - 0)$.

EXERCISES 2.4

1. There is a break in the graph, so it is not differentiable at $x = -1$.

3. There is a cusp at $x = 2$, so the function is not differentiable there.

5. It is differentiable at $x = 1/2$.

7. It is not continuous at $x = -1$ since there is a break in the graph there.

9. It is continuous at $x = 2$ since there is no break there.

11. It is continuous at $x = 1/2$ since there is no break there.

13. a) It is a polynomial and differentiable everywhere, including at $x = 0$.

b) It is a polynomial and continuous everywhere, including at $x = 0$.

15. a) The piecewise function has a cusp at $x = 0$, so it is not differentiable there.

b) The segments join at $x = 0$, so the function is continuous there.

17. a) The piecewise function has a cusp at $x = 0$, so it is not differentiable there.

b) The segments join at $x = 0$, so the function is continuous there.

19. a) It is a polynomial and differentiable everywhere, including at $x = 1$.

b) It is a polynomial and continuous everywhere, including at $x = 1$.

21. a) The piecewise function has a cusp at $x = 1$, so it is not differentiable there.

b) The segments join at $x = 1$, so the function is continuous there.

23. a) It is a polynomial and differentiable everywhere, including at $x = 2$.

b) It is a polynomial and continuous everywhere, including at $x = 2$.

25. a) The piecewise function has a cusp at $x = 2$, so it is not differentiable there.

b) The segments join at $x = 2$, so the function is continuous there.

27. For the function to be continuous for all x it must have a limit at exceptional values of x. The limit here is $\lim_{x \to -6} \frac{(x+6)(x-6)}{(x+6)} = \lim_{x \to -6} (x-6) = -12$. Any function that results in -12 when $x = -6$ is acceptable. The two simplest possibilities are a constant or the reduced function, $x - 6$. Therefore,

$$f(x) = \begin{cases} \dfrac{x^2 - 36}{x+6} & x \neq -6 \\ x - 6 & x = -6 \end{cases}$$

29. For the function to be continuous for all x it must first have a limit as x approaches 5. The limit is $\lim_{x \to 5} \frac{(x)(x-5)(x+5)}{(x-5)} = \lim_{x \to 5}(x)(x+5) = 50$. Any function value of 50 when x is 5 is acceptable. Therefore,

$$f(x) = \begin{cases} \dfrac{x^3 - 25x}{x-5} & x \neq 5 \\ x^2 + 5x & x = 5 \end{cases}$$

EXERCISES 2.5

1. $f'(x) = 4(4x^2 + 1)^3(8x) = 32x(4x^2 + 1)^3$

3. $y' = 7(5x^2 + 3)^6(10x) = 70x(5x^2 + 3)^6$

5. $f'(x) = 28(3x^2 + 1)^3(6x) = 168x(3x^2 + 1)^3$

7. $f'(x) = 5(9x^{10} + 6x^5 - x)^4(90x^9 + 30x^4 - 1)$

9. $y' = 10(12x^7 + 3x^4 - 2x + 5)^9(84x^6 + 12x^3 - 2)$

11. $f'(x) = \dfrac{-20}{(9x^3 + \sqrt{x} + 3)^6}\left(27x^2 + \dfrac{1}{2}x^{-1/2}\right)$

13. $f'(x) = 6(7x^{8/5} + 5x + 1)^5\left(\dfrac{56}{5}x^{3/5} + 5\right)$

15. $f'(x) = 5(\sqrt{x} + 1)^4\left(\dfrac{1}{2}x^{-1/2}\right)$

17. $y' = 6\left(4x^5 + \sqrt[3]{x^2} + 1\right)^5\left(20x^4 + \dfrac{2}{3}x^{-1/3}\right)$

19. $y' = 21x^2 + 16x - \dfrac{140}{(4x-3)^8}$

21. One seeks $y'(1)$. It is $2[5(1) + 1](5) = 60$

23. The point is (2, 27). The slope is $m = y'(2) = (3/2)[4(2) + 1]^{1/2}(4) = 18$. The equation of the tangent line is $y - 27 = 18(x - 2)$.

25. The point is (1, 5). The slope is $y'(1) = \dfrac{-15}{[(1)^3 - 2(1) + 2]^4}[3(1)^2 - 2]$ or -15. The equation of the tangent line is $y - 5 = -15(x - 1)$.

EXERCISES 2.6

1. $f'(z) = 4z^3 + 6z$

3. $\dfrac{ds}{dr} = 12r^3 + 3r^2 + 4r$

5. $f'(p) = 24p^7 + 30p^5 + 6p^2 + 4$

7. $f'(t) = 10(5t^2 + 3t + 1)^9(10t + 3)$

9. $S'(p) = 12(3p^{10} + \sqrt{p} + 5)^{11}\left(30p^9 + \dfrac{1}{2}p^{-1/2}\right)$

11. $\dfrac{d}{dt}(3t^{7/5} - 5t - 4) = \dfrac{21}{5}t^{2/5} - 5$

13. $\dfrac{d}{dp}(5p^4 - 3p^{2/3}) = 20p^3 - 2p^{-1/3}$

15. $\dfrac{d}{dt}(2a^7t^5 - 9bt^3 + t^2 + 3) = 10a^7t^4 - 27bt^2 + 2t$

$\dfrac{d}{db}(2a^7t^5 - 9bt^3 + t^2 + 3) = -9t^3$

17. $f'(x) = 27x^2 + 4$ and $f''(x) = 54x$

19. $y' = 6x^2 + 3 + \frac{3}{4}x^{-1/4}$ and $y'' = 12x - \frac{3}{16}x^{-5/4}$

21. $v' = 6t^2 + 18$ and $v'' = 12t$

23. The derivative is $3x^2 - 7$. Evaluating at $x = 1$ yields $3(1)^2 - 7 = -4$

25. The second derivative is $30t^4$. Evaluating at $t = -1$ yields 30.

27. The second derivative is $15(2x + 7)^{1/2}$. Evaluating at $x = 1$ yields 45.

29. $y' = 5x^4 + 9x^2 + 9$, $y'' = 20x^3 + 18x$, $y''' = 60x^2 + 18$, and $y^{iv} = 120x$.

EXERCISES 2.7

1. $\Delta f(x) = [6(x + 1) + 4] - [6x + 4] = 6x + 6 + 4 - 6x - 4 = 6$

3. $\Delta f(x) = [4(x + 1)^2] - [4x^2] = 4x^2 + 8x + 4 - 4x^2 = 8x + 4$

5. $\Delta f(x) = [5(x + 1)^2 + 2(x + 1) + 3] - [5x^2 + 2x + 3]$

$= 5x^2 + 10x + 5 + 2x + 2 + 3] - 5x^2 + 2x + 3] = 10x + 7$

7. $\Delta f(x) = [(x + 1)^3 + 3(x + 1) + 1] - [x^3 + 3x + 1]$

$= [x^3 + 3x^2 + 3x + 1 + 3x + 3 + 1] - [x^3 + 3x + 1]$

$= 3x^2 + 3x + 4$

9. $\Delta f(x) = [9(x + 1) - 1] - [9x - 1] = 9x + 9 - 1 - 9x + 1 = 9$ since $\Delta f(x)$ is a constant $\Delta^2 f(x) = 0$. Alternatively,

$\Delta^2 f(x) = f(x + 2) - 2f(x + 1) + f(x)$

$= [9(x + 2) - 1] - 2[9(x + 1) - 1] + [9x - 1] = 0.$

11. $\Delta^2 f(x) = [2(x + 2)^2 + 5] - 2[2(x + 1)^2 + 5] + [2x^2 + 5]$

$= 2x^2 + 8x + 8 + 5 - 4x^2 - 8x - 4 - 10 + 2x^2 + 5 = 4$

13. $\Delta^2 f(x) = [5(x + 2)^3 + 2] - 2[5(x + 1)^3 + 2] + [5x^3 + 2]$

$= 5x^3 + 30x^2 + 60x + 40 + 2$

$- 10x^3 - 30x^2 - 30x - 10 - 4 + 5x^3 + 2$

$= 30x + 30$

SUPPLEMENTARY EXERCISES CHAPTER 2

1. Evaluating the derivative at $x = 1/4$ yields $m = 2(1/4) = 1/2$.

3. When $2x = 3/4$. $x = 3/8$, and the point is $(3/8, 9/64)$.

5. $\lim_{x\to 2} 3x + \frac{4}{x} = 3(2) + \frac{4}{2} = 8$

7. This is an indeterminate form. Therefore,

$$\lim_{x\to 4} \frac{x^2 + 2x - 24}{x - 4} = \lim_{x\to 4} \frac{(x-4)(x+6)}{x-4} = 4 + 6 = 10$$

9. If $f(x) = 3x^2 + 5x + 1$

a) $$\lim_{h\to 0} \frac{[3(x+h)^2 + 5(x+h) + 1] - [3x^2 + 5x + 1]}{h}$$

$$= \lim_{h\to 0} \frac{6xh + 3h^2 + 5h}{h} = \lim_{h\to 0} \frac{h(6x + 3h + 5)}{h} = 6x + 5 = f'(x)$$

b) $$\lim_{x\to a} \frac{[3x^2 + 5x + 1] - [3a^2 + 5a + 1]}{x - a}$$

$$= \lim_{x\to a} \frac{(3x^2 - 3a^2) + (5x - 5a)}{x - a}$$

$$= \lim_{x\to a} \frac{3(x-a)(x+a) + 5(x-a)}{x - a} = 6a + 5 = f'(a).$$

Therefore, $f'(x) = 6x + 5$.

11. $f(1) = 5(1)^3 + 2(1)^2 + 3 = 10$ and $f'(1) = 15(1)^2 + 4(1) = 19$.

13. Evaluating $y' = 35x^6 - 6x$ at $x = 1$ yields a slope of 29. The point is $(1, 2)$ so the equation of the tangent line is $y - 2 = 29(x - 1)$.

15. First $f(1)$ must exist. Here, $f(1) = 2(1) - 3 = -1$. Next the limit at $x = 1$ must exist. So, $\lim_{x\to 1} \frac{(2x-3)(x-1)}{(x-1)} = 2(1) - 3 = -1$.
Since $f(1) = \lim_{x\to 1} f(x)$, the function is continuous.

17. $\frac{d}{dx}(9x^3 + 4x^2 + 3x + 1)^{25} = 25(9x^3 + 4x^2 + 3x + 1)^{24}(27x^2 + 8x + 3)$

19. $f'(x) = 4\left(2\sqrt[3]{x^2} + 3x + 1\right)^3 \left(\frac{4}{3}x^{-1/3} + 3\right)$

21. $y' = \frac{5}{2}(4x+1)^{3/2}(4) = 10(4x+1)^{3/2}$ at $x = 6$ the derivative is 1250. Therefore, $m = 1250$ and the point (6, 3125). The tangent line is $y - 3125 = 1250(x - 6)$.

23. $\frac{d}{dp}(5a^3p^4 + 3ap^2 + 2bp + c) = 20a^3p^3 + 6ap + 2b)$ and

$\frac{d}{da}(5a^3p^4 + 3ap^2 + 2bp + c) = 15a^2p^4 + 3p^2$

25. The first derivative is $36x^8 - 21x^6 + 7$ and the second derivative is $288x^7 - 126x^5$. Evaluating, $y''(1) = 288(1)^7 - 126(1)^5 = 162$

CHAPTER 3

USING THE DERIVATIVE

1. c and d are increasing for all x.

3. e has slope that is always increasing as x increases.

5. b, d, g, and h have an inflection point.

7. The graph is increasing on $(2, \infty)$ and decreasing on $(-\infty, 2)$. It has a local and absolute minimum at $(2, -1)$. The graph is concave up on $(-\infty, \infty)$ with no inflection points. The y-intercept is $(0, 3)$ and x-intercepts $(1, 0)$ and $(3, 0)$. There are no undefined points and no asymptotes.

9. The graph is decreasing on $(-\infty, -1) \cup (0, 1)$ and increasing on $(-1, 0) \cup (1, \infty)$. There is a local maximum at $(0, 0)$ and local and absolute minimums at $(\pm 1, -1)$. The graph is concave down on $(-3/4, 3/4)$ and concave up on $(-\infty, -3/4) \cup (3/4, \infty)$. There are inflection points at $(\pm 3/4, -1/2)$. The y-intercept is $(0, 0)$ and x-intercepts at $(0, 0)$, $(\pm 1.4, 0)$. There are no asymptotes and no undefined points.

11. There are endpoint extrema at A and G. Local extrema are at B, D, and F.

13. There are inflection points at C and E.

15. There is an absolute maximum at D and absolute minimum at F.

EXERCISES 3.2

1. A positive first derivative means that the function (graph) is increasing. Graphs c and d depict a positive first derivative for all x.

3. A positive second derivative means that the function (graph) is concave up. Graph e depicts a positive second derivative for all x.

5. One possibility is depicted in the graph as follows.

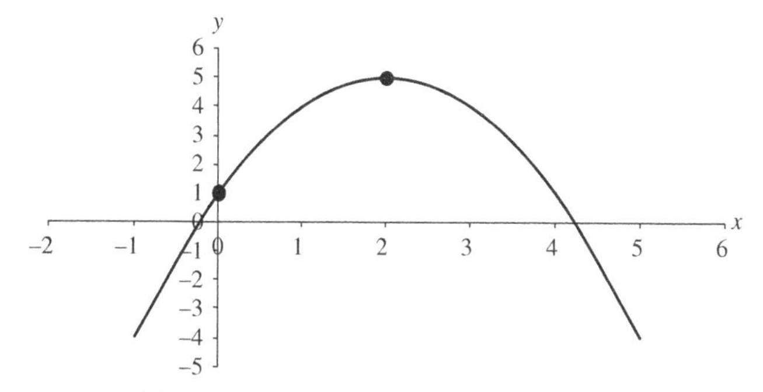

7. One possibility is depicted in the following graph.

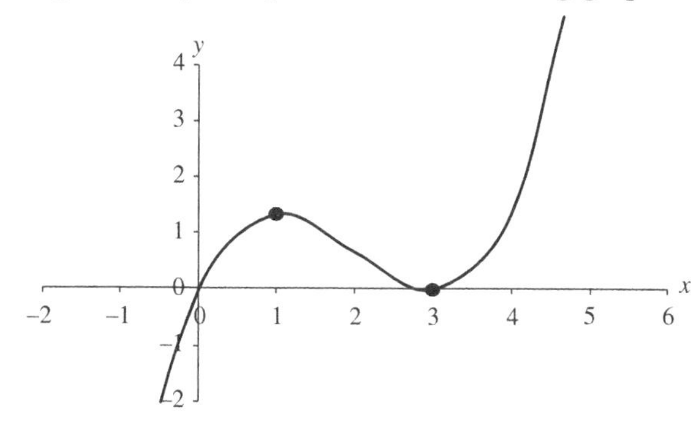

9. The table is filled in as follows.

	f	f'	f''
A	+	–	+
B	–	**0**	+
C	–	–	**0**
D	**0**	+	+

11. $f'(5)$ is positive since the graph is shown above the axis here. A positive first derivative indicates an increasing function.

13. A local maximum occurs when the derivative changes sign from positive to negative. This must occur at one of the x-intercepts of the derivative (see graph). The x-intercept is at $x = 1$.

15. The ordered pair is $(2, -2/3)$ and the slope is $f'(2) = -6$. The tangent line is $y + (2/3) = -6(x - 2)$.

EXERCISES 3.3

1. Setting the derivative to zero yields $2x - 4 = 0$ or $x = 2$ as a possible extremum. The derivative is negative on $(-\infty, 2)$ and positive on $(2, \infty)$ indicating a local minimum at $x = 2$. The minimum is $f(2) = -9$. Alternatively, the second derivative being positive at $x = 2$ also indicates a minimum there.

3. Setting the derivative to zero yields $3x^2 - 6x + 3 = 3(x - 1)^2 = 0$ or $x = 1$ as the only possible extremum. The derivative is positive on $(-\infty, 1)$ and remains positive on $(1, \infty)$. Since the first derivative does not change in sign, there is no extrema at $x = 1$. The function has no local extrema.

5. Using the first and second derivatives, $f'(x) = 2x + 8$ and $x = -4$ is the only possible extremum. The first derivative indicates that the function decreases on $(-\infty, -4)$ and increases on $(-4, \infty)$. The second derivative is $f''(x) = 2$ so, the graph is concave up everywhere and there is no inflection. Next, the y-intercept is 7 and x-intercepts are -1 and -7. At least five points are graphed as a check. The graph is:

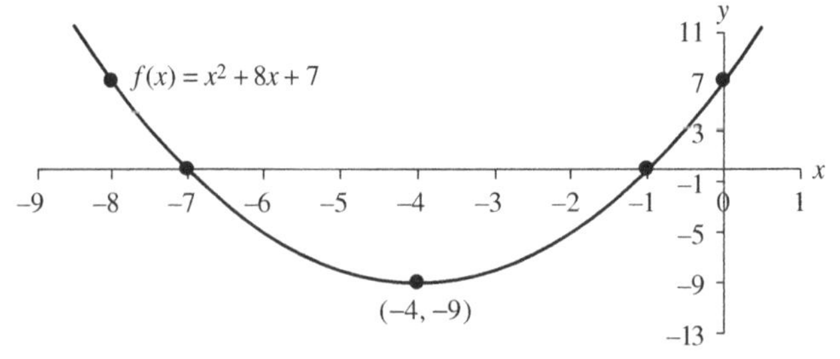

7. The first derivative $f'(x) = 3x^2 - 4$ indicates that the function is increasing on $\left(-\infty, \frac{-2\sqrt{3}}{3}\right) \cup \left(\frac{2\sqrt{3}}{3}, \infty\right)$ and decreasing on $\left(\frac{-2\sqrt{3}}{3}, \frac{2\sqrt{3}}{3}\right)$. There is a local minimum at $\left(\frac{2\sqrt{3}}{3}, \frac{-16\sqrt{3}}{9}\right)$

⟶

and a local maximum at $\left(\frac{-2\sqrt{3}}{3}, \frac{16\sqrt{3}}{9}\right)$. The second derivative $f''(x) = 6x$ indicates that the function is concave down on $(-\infty, 0)$ and concave up on $(0, \infty)$ with an inflection point at (0, 0). The function has x-intercepts at -2, 0, and 2, (from factoring) and the graph is

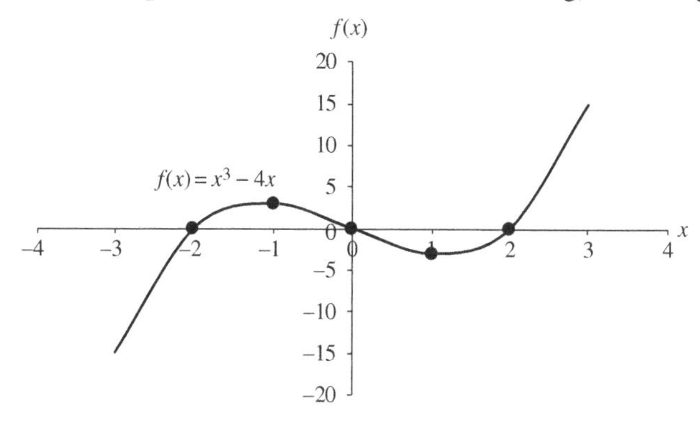

9. The first derivative $f'(x) = 12x^2 - 12x + 3$ indicates that the function is always increasing, so there are no local extrema. The second derivative $f''(x) = 24x - 12$ indicates that the function is concave down on $\left(-\infty, \frac{1}{2}\right)$ and concave up on $\left(\frac{1}{2}, \infty\right)$ with an inflection point at $\left(\frac{1}{2}, \frac{9}{2}\right)$. The function has a y-intercept of 4 and the graph is:

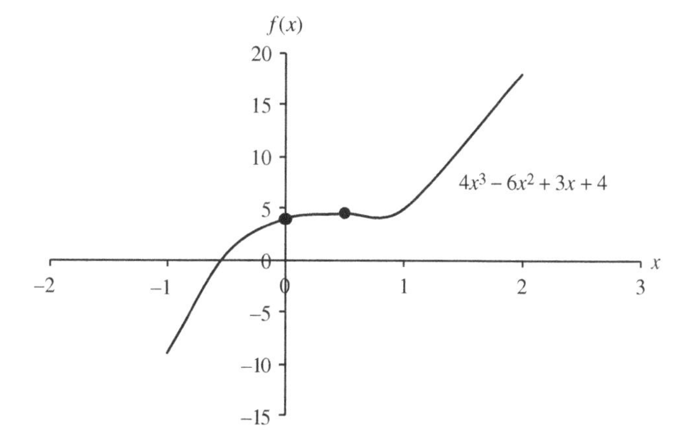

11. The first derivative $f'(x) = -3x^2 + 3x$ indicates that the function is increasing on (0, 1) and decreasing on $(-\infty, 0) \cup (1, \infty)$. There is a local minimum at $(0, 3)$ and a local maximum at $\left(1, \frac{7}{2}\right)$. The second

⟶

derivative, $f''(x) = -6x + 3$, indicates that the function is concave down on $\left(\frac{1}{2}, \infty\right)$ and concave up on $\left(-\infty, \frac{1}{2}\right)$ with an inflection point at $\left(\frac{1}{2}, \frac{13}{4}\right)$. The function has a y-intercept of 3 and the graph is

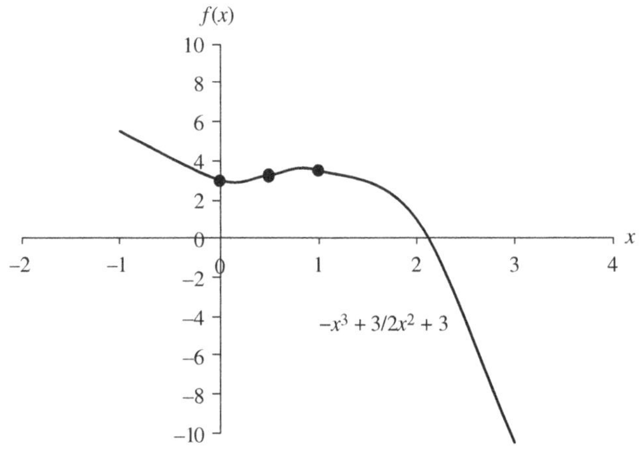

13. The first derivative $f'(x) = -x^2 - 4x + 1$ indicates that the function is increasing on $(-2 - \sqrt{5}, -2 + \sqrt{5})$ and decreasing on $(-\infty, -2 - \sqrt{5}) \cup (-2 + \sqrt{5}, \infty)$. Using two place decimal approximations, there is a local minimum at (−4.23, −8.79) and a local maximum at (0.23, 6.13). The second derivative, $f''(x) = -2x - 4$, indicates that the function is concave down on $(-2, \infty)$ and concave up on $(-\infty, -2)$ with an inflection point at $\left(-2, -\frac{4}{3}\right)$. The function has a y-intercept of 6 and the graph is:

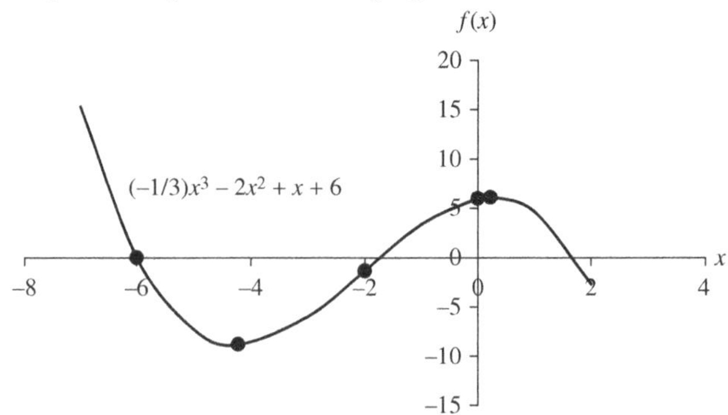

15. The first derivative $f'(x) = 4x^3 - 4x$ indicates that the function is decreasing on $(-\infty, -1) \cup (0, 1)$ and increasing on $(-1, 0) \cup (1, \infty)$. There are local minimums at $(\pm 1, -6)$ and a local maximum at $(0, -5)$. The second derivative $f''(x) = 12x^2 - 4$

indicates that the function is concave down on $\left(\frac{-\sqrt{3}}{3}, \frac{\sqrt{3}}{3}\right)$ and concave up on $\left(-\infty, \frac{-\sqrt{3}}{3}\right) \cup \left(\frac{\sqrt{3}}{3}, \infty\right)$ with inflection points at $\left(\frac{\pm\sqrt{3}}{3}, \frac{-50}{9}\right)$. The function has a y-intercept at the origin, x-intercepts at $x = \pm\sqrt{1+\sqrt{6}}$. The graph is:

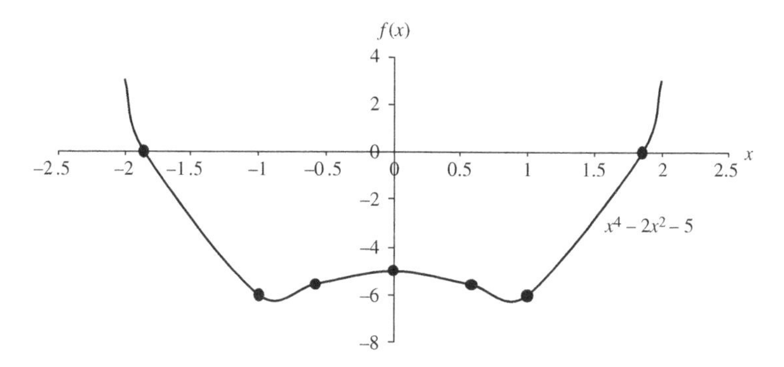

17. The first derivative $f'(x) = 1 - \frac{1}{x^2}$ indicates that the function is increasing on $(-\infty, -1) \cup (1, \infty)$ and decreasing on $(-1, 0) \cup (0, 1)$. The function is undefined at $x = 0$. There is a local minimum at $(1, 2)$ and a local maximum at $(-1, -2)$. The second derivative $f''(x) = \frac{2}{x^3}$ indicates that the function is concave down on $(-\infty, 0)$ and concave up on $(0, \infty)$ with no inflection points. There are asymptotes at $x = 0$ and at $y = x$. The function has no y-intercept, and the graph is:

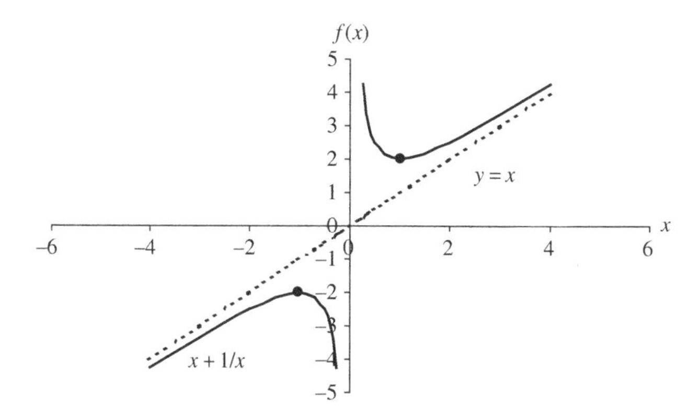

EXERCISES 3.4

1. Setting the derivative to zero yields $f'(x) = 2x - 6 = 0$. Therefore, at $x = 3$, there is a possible extremum. The derivative is negative when $x < 3$ and positive when $x > 3$ indicating a local minimum at $x = 3$. (Alternatively, the positive second derivative indicates a minimum at the critical value $x = 3$ from the first derivative.)

3. $f'(x) = -4x^3 + 6x$

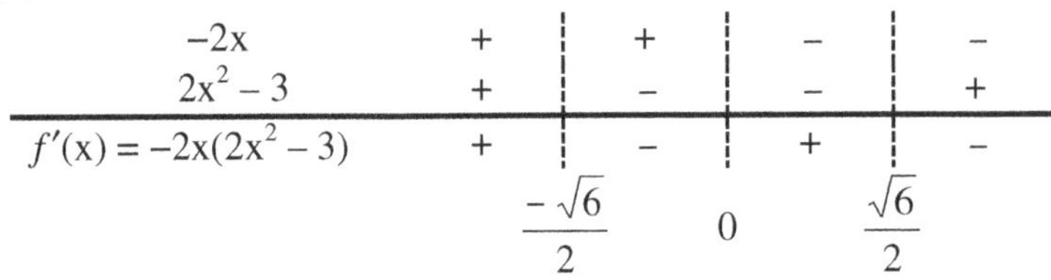

$-2x$	+	+	−	−
$2x^2 - 3$	+	−	−	+
$f'(x) = -2x(2x^2 - 3)$	+	−	+	−

Critical points: $-\frac{\sqrt{6}}{2}$, 0, $\frac{\sqrt{6}}{2}$

The critical point table indicates local maxima at $x = \frac{\pm\sqrt{6}}{2}$. The maximum value is 9/4. Also, negative signed second derivatives at the critical values (first derivative) signal a maximum.

5. $f'(x) = 16 - 2x$. The derivative is positive when $x < 8$ and negative when $x > 8$ indicating a local maximum at $x = 8$. The maximum value is 64. Alternatively, the negative second derivative indicates a maximum at the critical value $x = 8$.

7. $x + y = 8$ and $y = 8 - x$ so the product is $P(x) = x(8 - x) = 8x - x^2$. $P'(x) = 8 - 2x$ indicates a possible extremum at $x = 4$. The derivative is positive when $x < 4$ and negative when $x > 4$ indicating a maximum at $x = 4$. The maximum of the product is 16. Alternatively, the negative second derivative indicates a maximum at the critical value $x = 4$.

9. $x + y = 20$ and $y = 20 - x$ so the sum of the squares, $S(x) = x^2 + (20 - x)^2 = 2x^2 - 40x + 400$. $S'(x) = 4x - 40$, indicates a possible extremum at $x = 10$. The derivative is negative when $x < 10$ and positive when $x > 10$ indicating a minimum at $x = 10$. The minimum sum of squares is 200. Alternatively, the positive second derivative indicates a minimum at the critical value $x = 10$.

11. First, for a number to exceed its square, it is between 0 and 1. One seeks to maximize $f(x) = x - x^2$. The first derivative $f'(x) = 1 - 2x$ indicates a possible extremum at $x = 1/2$. The derivative is positive when $x < 1/2$ and negative when $x > 1/2$. Therefore, 1/2 exceeds its square by the largest amount. Alternatively, the negative second derivative also indicates a maximum at the critical value $x = 1/2$.

13. Here, one seeks to maximize the area xy subject to $2x + 2y = 60$. Solving for y in the constraint yields $y = 30 - x$. Therefore, one seeks to maximize $A(x) = x(30 - x) = 30x - x^2$. The first derivative, $A'(x) = 30 - 2x$, indicates a possible extremum when $x = 15$. The first derivative is positive when $x < 15$ and negative when $x > 15$ indicating a maximum at $x = 15$. Therefore, $x = y = 15$. The maximum area is a square of 225 ft^2. Alternatively, the negative second derivative also indicates a maximum at the critical value $x = 15$.

15. Here, one seeks to maximize $xy/2$ subject to the constraint $x + y = 10$. The objective written in terms of a single variable is $A(x) = \frac{1}{2}(x)(10 - x) = \frac{-1}{2}x^2 + 5x$. The first derivative $A'(x) = -x + 5$ indicates a possible extremum at $x = 5$. The derivative is positive when $x < 5$ and negative when $x > 5$, so there is a maximum at $x = 5$. Alternatively, the second derivative being negative at the critical value $x = 5$ also indicates a maximum there. If $x = 5$ then y is also 5, so the largest area for the right triangle is 25/2 square centimeters.

17. First, rewrite the constraint as $y = 8 - x$. The objective function to be minimized is $S(x) = x^2 + (8 - x)^2 = 2x^2 - 16x + 64$. $S'(x) = 4x - 16$ indicates a possible extremum at $x = 4$. When $x < 4$, the first derivative is negative and when $x > 4$, the first derivative is positive. There is a minimum at $x = 4$. Alternatively, the second derivative being positive at the critical value $x = 4$ also indicates a minimum there. The minimum value for the sum of squares is 32.

19. The objective is to maximize the area xy subject to the constraint $12x + 10y = 120$. Solving for y yields $y = 12 - \frac{6}{5}x$. Rewriting the objective as $A(x) = x\left(12 - \frac{6}{5}x\right) = 12x - \frac{6}{5}x^2$ and taking the derivative yields $A'(x) = 12 - \frac{12}{5}x$. There is a possible extremum at $x = 5$. The derivative is positive when $0 < x < 5$ and negative when $x > 5$ indicating a local maximum at $x = 5$. Alternatively, the second derivative being negative at the critical value $x = 5$ indicates a maximum there. The maximum area is 30 ft^2.

21. To maximize xy subject to the constraint $15x + 20y = 2400$, solve for y and substitute into the objective function to yield $A(x) = x\left(120 - \frac{3}{4}x\right) = 120x - \frac{3}{4}x^2$. Taking the first derivative yields $A'(x) = 120 - \frac{3}{2}x$ indicating a possible extremum at $x = 80$. The derivative is positive when $0 < x < 80$ and negative when $x > 80$ indicating a maximum at $x = 80$ ft and $y = 60$ ft. Alternatively, the

⟶

second derivative being negative at the critical value $x = 80$ indicates a maximum there.

23. To maximize x^2h subject to the constraint $2x^2 + 4xh = 60$, solve for y and substitute to yield an objective function $V(x) = x^2\left(\frac{15}{x} - \frac{1}{2}x\right) = 15x - \frac{1}{2}x^3$ and a first derivative of $V'(x) = 15 - \frac{3}{2}x^2$. The derivative indicates a possible extremum at $x = \sqrt{10}$ (no negative lengths). The first derivative is positive on $0 < x < \sqrt{10}$ and negative when $x > \sqrt{10}$ indicating a local maximum at $x = h = \sqrt{10}$ ft. The box is a cube with dimensions $\sqrt{10} \times \sqrt{10} \times \sqrt{10}$. Alternatively, the second derivative being negative at the critical value indicates a maximum there.

25. Here, the surface area is $2\pi rh + 2(2r)^2 = 2\pi rh + 8r^2$ and the volume is $V = \pi r^2 h$. Using the volume and solving for h yields $h = \frac{V}{\pi r^2}$. The objective is to minimize $S(r) = 2\pi r\left(\frac{V}{\pi r^2}\right) + 8r^2 = \frac{2V}{r} + 8r^2$. The first derivative is $S'(r) = \frac{-2V}{r^2} + 16r$ indicates a possible extremum at $r = \frac{V^{1/3}}{2}$. For a minimum, $r = \frac{V^{1/3}}{2}$ and $h = \frac{4V^{1/3}}{\pi}$. The optimal height to diameter ratio is $\frac{h}{2r} = \frac{4V^{1/3}}{\pi(V^{1/3})} = \frac{4}{\pi}$. This relationship holds regardless of the volume, V, and indicates that the can is slightly taller than in Example 3.4.8.

27. Let c = cost per square foot of the sides and $2c$ = cost per square foot for the top and bottom. The area of the bottom is $\pi\left(\frac{d}{2}\right)^2 = \frac{\pi d^2}{4}$, where d is the diameter of the circular end. The volume of the cylinder is $V = \frac{\pi d^2}{4}(h)$. Solving for h yields $h = \frac{4V}{\pi d^2}$. The cost as a function of d is $C(d) = 2\left(\frac{\pi d^2}{4}\right)(2c) + \pi d\left(\frac{4V}{\pi d^2}\right)(c) = \pi d^2 c + \frac{4cV}{d}$. The first derivative is $C'(d) = 2\pi cd - \frac{4cV}{d^2}$ indicating a possible extremum when $d = \sqrt[3]{\frac{2V}{\pi}}$. The minimum occurs when $d = \sqrt[3]{\frac{2V}{\pi}}$ and $h = \sqrt[3]{\frac{16V}{\pi}}$, so the best height to diameter ratio $\frac{h}{d} = 2$.

EXERCISES 3.5

1. $C'(x) = 6x^2 + 9$

3. $C'(x) = 7x^6 - 15x^2 + 20$ so $C'(1) = 7(1)^6 - 15(1)^2 + 20 = \$12$

5. $\text{MC} = C'(x) = 3x^2 - 18x + 15 = 3(x-1)(x-5)$. Setting $\text{MC} = 0$ yields $x = 1$ or 5.

7. $\text{MC} = C'(x) = 3x^2 - 24x + 60$. Next, the second derivative is needed to find a possible minimum. $\text{MC}'(x) = 6x - 24$. Setting this derivative to zero yields $x = 4$, a local minimum. The minimum marginal cost is $\text{MC}(4) = C'(4) = 3(4)^2 - 24(4) + 60 = 12$.

9. $R'(x) = -2x + 30$

11. Revenue is price multiplied by demand.
So, $R(x) = xp(x) = x\left(-\dfrac{1}{4x} + 12\right) = \dfrac{-x^2}{4} + 12x$.
Therefore, $R'(x) = \dfrac{-x}{2} + 12$.

13. The restrictions for x and p are $0 \le x \le 40$ and $0 \le p \le 240$, respectively. Next, solving for p in terms of x yields $p = 240 - 6x$. Using this information, the revenue function is formed as $R(x) = (x)(240 - 6x) = 240x - 6x^2$. Using Profit = revenue − cost yields $\Pr(x) = (240x - 6x^2) - (x^3 + 9x^2 - 360x + 2000)$ $= -x^3 - 15x^2 + 600x - 2000$. The first derivative yields $Pr'(x) - 3x^2 - 30x + 600$. Setting it to zero yields $x = 10$ as a possible extremum (−20 is rejected). Other possibilities are $x = 0$ and $x = 240$ as endpoint extrema. Evaluating the profit at the potential candidates yields the maximum profit of \$1500 when 10 units are produced at \$180 each.

SUPPLEMENTARY EXERCISES CHAPTER 3

1. c and e are increasing for all x.

3. a and c are concave up for all x.

5. Here, $y' = \dfrac{-500}{(x+1)^2}$ since the denominator is always positive y' is always negative indicating that the function is decreasing for $x \ge 0$.

7. The first derivative is $f'(x) = 4x - 1$ and $x = 1/4$ is the only possible extremum. The first derivative indicates that the function decreases on $(-\infty, 1/4)$ and increases on $(1/4, \infty)$ so the minimum value is $-49/8$ when $x = 1/4$. The second derivative $f''(x) = 4$ indicates that the

graph is concave up everywhere and has no points of inflection. Next, there is a y-intercept of -6 and x-intercepts of 2 and $-3/2$.
The graph is:

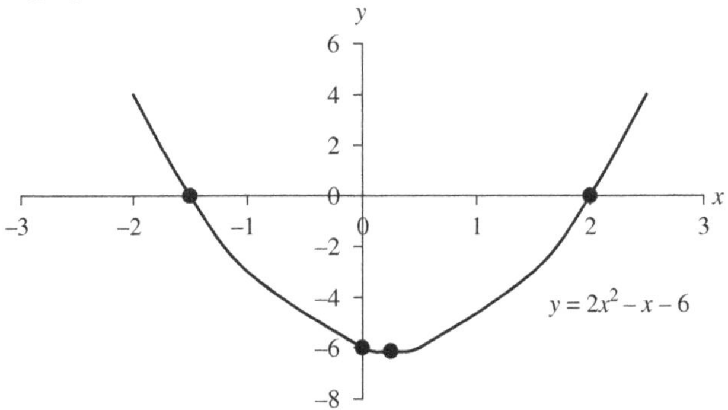

9. The first derivative $f'(x) = 4x^3 - 18x$ indicates that the function is decreasing on $\left(-\infty, \frac{-3\sqrt{2}}{2}\right) \cup \left(0, \frac{3\sqrt{2}}{2}\right)$ and increasing on $\left(-\frac{3\sqrt{2}}{2}, 0\right) \cup \left(\frac{3\sqrt{2}}{2}, \infty\right)$. There are local minimums at $\left(\frac{\pm 3\sqrt{2}}{2}, \frac{-81}{4}\right)$ and a local maximum at (0, 0). The second derivative $f''(x) = 12x^2 - 18$ indicates that the function is concave down on $\left(\frac{-\sqrt{6}}{2}, \frac{\sqrt{6}}{2}\right)$ and concave up on $\left(-\infty, \frac{-\sqrt{6}}{2}\right) \cup \left(\frac{\sqrt{6}}{2}, \infty\right)$ with inflection points at $\left(\frac{\pm\sqrt{6}}{2}, \frac{-45}{4}\right)$. The function has a y-intercept at the origin, x-intercepts at $x = -3, 0$, and 3. The graph is:

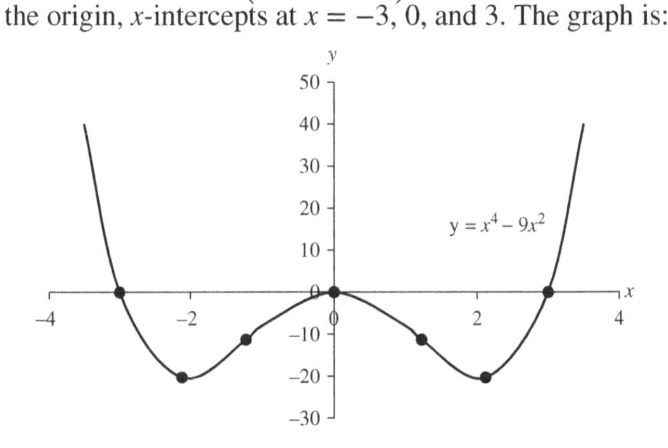

11. Here, $f'(x) = 12x^3 - 12x^2$ and $f''(x) = 36x^2 - 24x = 12x(3x - 2)$.

12x	–	+	+
3x – 2	–	–	+
f″(x) = 12x(3x – 2)	+	–	+
	0	2/3	

The critical point table shows that there are inflection points at $x = \frac{2}{3}$ and at $x = 0$. The inflection points are (0, 0) and $\left(\frac{2}{3}, \frac{-16}{27}\right)$.

13. To maximize the area $A(x) = xy$ subject to the cost constraint $20x + 30y = 480$, solve for y and substitute to yield
$A(x) = x\left(16 - \frac{2}{3}x\right) = 16x - \frac{2}{3}x^2$.
The first derivative is $A'(x) = 16 - \frac{4}{3}x$. When $x < 12$, the first derivative is positive and when $x > 12$, it is negative indicating a maximum occurs at $x = 12$ yards. The maximum area is 96 square yards when $x = 12$ yards and $y = 8$ yards. Alternatively, the second derivative being negative also indicates a maximum at the critical value $x = 12$. (Note as a quick check, the same sum is spent in each direction. In this case, $12(20) = 8(30) = 240$ dollars in each of the two directions x and y.)

15. a) In this case, one seeks to minimize $2x^2 + 4xh$ subject to $x^2h = 64$. Solving for y and substituting yields an objective function $S(x) = 2x^2 + 4x\left(\frac{64}{x^2}\right) = 2x^2 + \frac{256}{x}$ and a first derivative of $S'(x) = 4x - \frac{256}{x^2}$. The derivative indicates a possible extremum at $x = 4$. The first derivative is negative on $0 < x < 4$ and positive when $x > 4$ indicating a local minimum at $x = h = 4$ in. Alternatively, the second derivative being positive at the critical value $x = 4$ also indicates a minimum there. The box is a 4-inch cube.

b) Here, since the box is to be open, minimize $x^2 + 4xh$ subject to $x^2h = 64$. Solving for y and substituting yields an objective function $S(x) = x^2 + 4x\left(\frac{64}{x^2}\right) = x^2 + \frac{256}{x}$ and a first derivative of $S'(x) = 2x - \frac{256}{x^2}$. The derivative indicates a possible extremum at $x = 4\sqrt[3]{2}$. The first derivative is negative on $0 < x < 4\sqrt[3]{2}$ and

⟶

positive when $x > 4\sqrt[3]{2}$ indicating a local minimum at $x = 4\sqrt[3]{2}$ in. and $h = 2\sqrt[3]{2}$ in. Alternatively, the second derivative being positive at the critical value also indicates a minimum there. The box has a square base whose side is twice the height.

17. Here, one seeks to minimize $V(x) = \frac{c_1^2}{x_1} + \frac{c_2^2}{x_2}$ subject to $c_1x_1 + c_2x_2 = B$. Using the constraint one determines $x_2 = \frac{B - c_1x_1}{c_2}$ so one rewrites the objective as $V(x) = \frac{c_1^2}{x_1} + \frac{c_2^2}{\frac{B - c_1x_1}{c_2}} = \frac{c_1^2}{x_1} + \frac{c_2^3}{B - c_1x_1}$. Differentiating yields $V'(x) = \frac{-c_1^2}{x_1^2} + \frac{c_2^3c_1}{(B - c_1x_1)^2}$. Setting the derivative to zero yields the following $(c_1^3 - c_2^3)x_1^2 - 2c_1^2Bx_1 + c_1B^2 = 0$. Using the quadratic formula yields a possible optimum when $x_1 = \frac{c_1^2B \pm c_2B\sqrt{c_1c_2}}{c_1^3 - c_2^3}$

19. Here, $R'(x) = 4000 - 4x$ indicates an extremum at $x = 100$. At this point, the maximum revenue is \$2,000,000.

21. The restrictions for x and p are $0 \leq x \leq 40$ and $0 \leq p \leq 240$. Next, solving for p in terms of x yields $p = 240 - 6x$. Using this information, the revenue function is formed as $R(x) = (x)(240 - 6x) = 240x - 6x^2$. Using Profit = revenue − cost yields $\Pr(x) = (240x - 6x^2) - (x^3 - 21x^2 - 360x + 8350)$ $= -x^3 + 15x^2 + 600x - 8350$. Now, the first derivative yields $Pr'(x) = -3x^2 + 30x + 600$. Setting the derivative to zero yields 20 as a possible extremum (-10 is rejected as a possibility). Other possibilities are $x = 0$ and $x = 240$ as endpoint extrema. Evaluating the profit at the potential extrema yields the maximum profit of \$1650 when 20 units are produced at \$120 each.

CHAPTER 4

EXPONENTIAL AND LOGARITHMIC FUNCTIONS

EXERCISES 4.1

1. a) The expression must have powers of 2 or 3. In this case, 8 is a power of 2, so, $(8)^{3x} = (2^3)^{3x} = 2^{9x}$.

b) Rewrite as $(27)^{2x} = (3^3)^{2x} = 3^{6x}$.

c) Rewrite as $(16)^{5x} = (2^4)^{5x} = 2^{20x}$.

3. a) $\left(\frac{1}{8}\right)^{-4x} = \left(\frac{1}{2^3}\right)^{-4x} = (2^{-3})^{-4x} = (2)^{12x}$

b) $\left(\frac{1}{9}\right)^{6x} = \left(\frac{1}{3^2}\right)^{6x} = (3^{-2})^{6x} = (3)^{-12x}$

c) $\left(\frac{1}{27}\right)^{-2x} = \left(\frac{1}{3^3}\right)^{-2x} = (3^{-3})^{-2x} = (3)^{6x}$

5. a) $\dfrac{10^{5x}}{5^{5x}} = (2)^{5x}$

b) $\dfrac{32^{2x}}{16^{2x}} = (2)^{2x}$

c) $\dfrac{4^{3x}}{12^{3x}} = \left(\frac{1}{3}\right)^{3x} = (3^{-1})^{3x} = (3)^{-3x}$

Solutions Manual to Accompany Fundamentals of Calculus, First Edition. Carla C. Morris and Robert M. Stark.

7. $\dfrac{7x^3x^5y^{-2}}{x^2y^4} = \dfrac{7x^8}{x^2y^4y^2} = \dfrac{7x^6}{y^6}$

9. $\dfrac{x^3}{y^{-2}} \div \dfrac{x}{y^5} = (x^3y^2) \cdot \left(\dfrac{y^5}{x}\right) = x^2y^7$

11. $\dfrac{2^{5x+3}4^{x+1}}{8(2^{3x-1})} = \dfrac{2^{5x+3}(2^2)^{x+1}}{2^3(2^{3x-1})} = \dfrac{2^{5x+3}2^{2x+2}}{2^{3x+2}} = 2^{4x+3}$

13. Since the bases are the same, exponents are equal. Therefore, $3x = 15$ yields $x = 5$.

15. Equate bases as $2^{7-x} = 2^5$ to yield $7 - x = 5$ and $x = 2$.

17. Factoring yields $5^x[(1 + x) + (3-2x)] = 0$

$$5^x[(4-x)] = 0$$

Since 5^x cannot be zero, $4 - x = 0$ and $x = 4$.

19. Factoring, $7^x[(x^2 + 4x) + (x + 6)] = 0$. Simplifying, $7^x[x^2 + 5x + 6] = 0$. Therefore, $[x^2 + 5x + 6] = (x + 3)(x + 2) = 0$. The solutions are $x = -2$ or $x = -3$.

21. $2^{3+h} = 2^h(2^3)$

23. $7^{x+5} - 7^{2x} = 7^{2x}(7^{-x+5} - 1)$

25. This is a difference of cubes $(7^h)^3 - 8 = (7^h-2)[7^{2h} + 2(7^h) + 4]$.

EXERCISES 4.2

1. Rewrite as $3^y = x$ to determine ordered pairs and graph as:

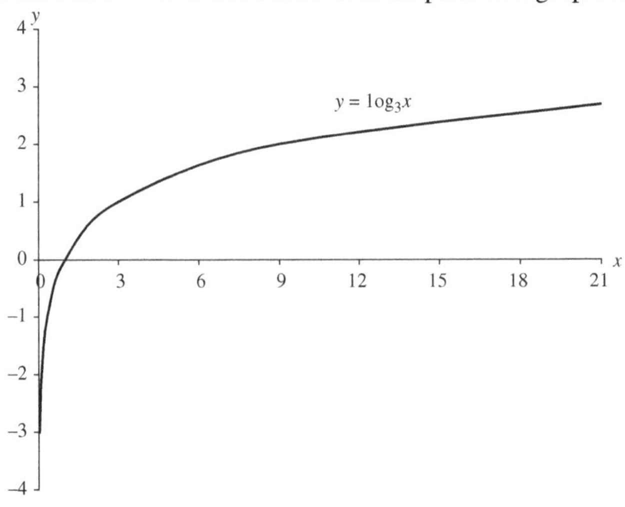

3. $\log_{10}1{,}000{,}000 = x$ so $10^x = 10^6$, which yields $x = 6$.

5. $\log_2 64 = x$ so $2^x = 2^6$, which yields $x = 6$.

7. $\log_2 \frac{1}{32} = x$ so $2^x = 2^{-5}$, which yields $x = -5$.

9. Since $\ln e^x = x$, then $\ln e^3 = 3$.

11. Since $\ln e^x = x$, then $\ln e^{7.65} = 7.65$.

13. $\ln(\ln e) = \ln(1) = 0$.

15. $\log_9 27 = x$ so $9^x = 27$, which yields $3^{2x} = 3^3$. Therefore, $2x = 3$ or $x = 3/2$.

17. $\log_4 32 = x$ so $4^x = 2^5$. This is rewritten as $2^{2x} = 2^5$, so $2x = 5$ and $x = 5/2$.

19. $\log_2 128 = x$ so $2^x = 128 = 2^7$. Therefore, $x = 7$.

21. Rewriting $\log_x 27 = 3$ in exponential form yields $x^3 = 27$ and $x = 3$.

23. Rewriting $\log_3(5x + 2) = 3$ in exponential form yields $3^3 = 5x + 2$. Solving for x yields $x = 5$.

25. Here, $5x = 35$ or $x = 7$.

27. Since $\ln(\ln 4x) = 0$, $\ln 4x = 1$. Therefore, $4x = e$ and $x = e/4$.

29.
$$\log_4 \frac{(x+1)^2(x-3)^6}{(3x+5)^3} = \log_4(x+1)^2 + \log_4(x-3)^6 - \log_4(3x+5)^3$$
$$= 2\log_4(x+1) + 6\log_4(x-3) - 3\log_4(3x+5).$$

31. First, the coefficients become exponents to yield $\ln x^2 - \ln(y+1)^3 + \ln(z+1)^4$.

This results in $\ln \dfrac{x^2(z+1)^4}{(y+1)^3}$.

33. Using logarithms, $10^{2x-1} = 105$ is written as $\log_{10}10^{2x-1} = \log_{10}105$, which yields $2x - 1 = \log_{10}105$. Therefore, $x = \dfrac{1 + \log_{10}105}{2}$.

35. First, rewrite as $e^{x-1} = 4/3$. Taking the natural logarithm of both sides of the equation yields $\ln e^{x-1} = \ln(4/3)$ or $x - 1 = \ln 4/3$. Therefore, $x = 1 + \ln(4/3)$.

37. The difference between 4.1 and 6.8 quakes is 2.7 on the Richter scale. Therefore, one seeks $10^{2.7} = 501.2$. The "Halloween" Quake was roughly 500 times stronger than the Northern Illinois Quake.

EXERCISES 4.3

1. Multiplying exponents yields e^{3x}, rewriting with a negative exponent yields e^{-5x}.

3. Add exponents to yield e^{7x}, $(e^{10x})^{1/2} = e^{5x}$.

5. Since e^2 is a constant then $f'(x) = 0$.

7. $f'(x) = 2e^{2x}$

9. $f'(x) = 5(6e^{6x}) = 30e^{6x}$

11. $f'(x) = 5(25e^{25x}) = 125e^{25x}$

13. $f'(x) = 6e^{8x^9}[72x^8] = 432x^8e^{8x^9}$

15. $f'(x) = e^{2x^3+5x+1}[6x^2 + 5]$

17. $f'(x) = 6e^{8x^7+9x^4}[56x^6 + 36x^3]$

19. Evaluating y and its derivative at $x = 0$ gives the point (0, 2) and $y'(0) = 5e^{5(0)} + 3 = 8 = m$. The tangent line is $y - 2 = 8(x - 0)$ or $y = 8x + 2$.

EXERCISES 4.4

1. $f'(x) = \dfrac{25}{25x} = \dfrac{1}{x}$

3. $f'(x) = 3x^2 + 10x + \dfrac{3}{3x} = 3x^2 + 10x + \dfrac{1}{x}$

5. $f'(x) = \dfrac{4}{4x + 7}$

7. $f'(x) = \dfrac{15x^2 + 12x + 9}{5x^3 + 6x^2 + 9x + 2}$

9. $f'(x) = \dfrac{45x^4 + 24x^3 + 8}{9x^5 + 6x^4 + 8x}$

11. $f'(x) = \dfrac{(1/2)(5x + 1)^{-1/2}(5)}{\sqrt{5x + 1}} = \dfrac{5}{2(5x + 1)}$

13. $f'(x) = 8(17x^2 + 15x + \ln 2x)^7 \left(34x + 15 + \dfrac{1}{x}\right)$

15. $f'(x) = \dfrac{3e^{3x}}{e^{3x} + 2}$

17. The point is $(e, 1)$ and the derivative at $x = 1$ is $1/e$, so the tangent line is $y - 1 = (1/e)(x - e)$.

19. Marginal cost is the derivative of the cost function, so
$MC = C'(x) = \dfrac{300}{3x+1}$.

21. Rewriting as $5\ln(3x+5) + 2\ln(4x^3 - 7x + 1)$ yields
$f'(x) = 5\left(\dfrac{3}{3x+5}\right) + 2\left(\dfrac{12x^2-7}{4x^3-7x+1}\right)$.

EXERCISES 4.5

1. Using $t = 5$ to represent 2015 yields $50{,}000e^{0.3(5)} \approx 58{,}092$.

3. There are seven doublings, so $500(2)^7 = 64{,}000$.

5. The formula for doubling is $250(1.06)^t = 500$. Solving for t yields the following

$$(1.06)^t = 2$$

$$t\ln(1.06) = \ln 2$$

$$t = \frac{\ln 2}{\ln 1.06} = 11.896 \approx 12\,\text{days}$$

7. The half-life is $P_0e^{-\lambda(28)} = 0.50P_0$. Therefore, $\lambda = 0.02476$. If there are 500 g today, in 200 years there will be $500e^{-0.02476(200)} \approx 3.54g$.

9. The half-life is $P_0e^{-\lambda(35)} = 0.50P_0$. Therefore, $\lambda = 0.0198$. One seeks the time to decay to 1/5 of the original. So, $P_0e^{-0.0198t} = \dfrac{1}{5}P_0$. Solving for t yields $t \approx 81.3$ days.

11. Using the model in the text, one has $2.5e^{-(0.00012097(10{,}000)} = 0.7457g$.

EXERCISES 4.6

1. The compound interest formula yields
$50{,}000\left(1 + \dfrac{0.08}{4}\right)^{4\cdot 4} = 50{,}000(1.02)^{16} = \$68{,}639.29$.

3. The continuous compound interest formula yields
$500{,}000e^{(0.07)(6)} = 500{,}000e^{0.42} = \$760{,}980.78$.

5. Comparing,
$250{,}000(1 + 0.03)^5 = 250{,}000(1.03)^5 = \$289{,}818.52$ and
$225{,}000e^{(0.05)(4)} = 225{,}000e^{0.20} = \$274{,}815.62$. Therefore, \$250,000 is worth more.

7. Solve $25{,}000e^{0.04t} = \$100{,}000$ so $e^{0.04t} = 4$ and $t = \dfrac{\ln 4}{0.04} \approx 34.66$ years.

9. $S_0 e^{-0.2t} = 0.3S_0$ and therefore, $t = \dfrac{\ln(0.3)}{-0.2} \approx 6.02$ weeks.

11. $\dfrac{f'(t)}{f(t)} = \dfrac{6t+2}{3t^2+2t+1}$. When $t = 3$, the percentage rate of change is 58.82%.

13. $\dfrac{f'(t)}{f(t)} = \dfrac{0.4e^{0.4t}}{e^{0.4t}} = 0.4$. When $t = 5$, the percentage rate of change is 40%. This rate is independent of t.

15. $\dfrac{f'(t)}{f(t)} = \dfrac{\dfrac{-1}{(t+3)^2}}{\dfrac{1}{t+3}} = \dfrac{-1}{t+3}$. When $t = 2$, the percentage rate of change is -20%.

17. For $E(p) = \dfrac{-p(-500e^{-0.25p}(0.25)}{500e^{-0.25p}} > 1$, $0.25p > 1$ and $p > 4$.

SUPPLEMENTARY EXERCISES CHAPTER 4

1. a) $\dfrac{16^{3x}}{8^{3x}} = \left(\dfrac{16}{8}\right)^{3x} = 2^{3x}$ b) $\dfrac{50^{5x-1}}{10^{5x-1}} = \left(\dfrac{50}{10}\right)^{5x-1} = 5^{5x-1}$

c) $\dfrac{25^{2x+1}5^{3x+2}}{125^x} = \dfrac{(5^2)^{2x+1}5^{3x+2}}{(5^3)^x} = \dfrac{5^{4x+2}5^{3x+2}}{5^{3x}} = 5^{4x+4}$

3. a) Rewrite as $5^{-2x} = 5^4$ so $x = -2$.

b) Rewrite as $2^{3x-1} = 2^5$ so $3x - 1 = 5$ and $x = 2$.

5. This is a difference of squares and factors as $(3^x - 5)(3^x + 5)$.

7. $f'(x) = 4(9e^{9x}) = 36e^{9x}$

9. $f'(x) = 12x^2(e^{4x^3})$

11. $f'(x) = 6e^{6x}$ and at $x = 0$, one has $m = f'(0) = 6$. Since $f(0) = 2$, one seeks a line through $(0, 2)$ with slope 6 so $y - 2 = 6(x - 0)$ or $y = 6x + 2$.

13. Solve by setting $3^x = 729$. Then $3^x = 3^6$ and $x = 6$. Therefore, $\log_3 729 = 6$.

15. Solve by setting $9^x = 27$. Then $3^{2x} = 3^3$ so $2x = 3$ and $x = 3/2$. Therefore, $\log_9 27 = 3/2$.

17. $\ln e^{7.23} = 7.23$ since $\ln e^x = x$.

19. Using rules for logarithms,
$\ln (2x+3)^4(x+1)^2(4x+5)^7$
$= \ln (2x+3)^4 + \ln (x+1)^2 + \ln (4x+5)^7$
$= 4\ln(2x+3) + 2\ln(x+1) + 7\ln(4x+5)$.

21. The earthquake is $10^{8.9-6.3} = 10^{1.4} \approx 25.1$ times as strong.

23. $f'(x) = \left[\dfrac{15x^2+4x+8}{5x^3+2x^2+8x+7}\right]$

25. $650 = 250e^{\lambda(5)}$ so $\lambda = 0.1911$. When $t = 10$, one has $250e^{(0.1911)(10)} \approx 1690$.

27. Solving $P_0 e^{\lambda(1)} = 3P_0$ yields $\lambda = \ln 3$, so $400e^{(\ln 3)(5)} = 97{,}200$. An alternative method uses $400(3)^5 = 97{,}200$.

29. $\dfrac{f'(t)}{f(t)} = \dfrac{-0.02e^{-t} + 0.03}{0.3 + 0.03t + 0.02e^{-t}}$

When $t = 0$, the percentage rate of change is 3.125%.
When $t = 2$, the percentage rate of change is 7.53%.

CHAPTER 5

TECHNIQUES OF DIFFERENTIATION

EXERCISES 5.1

1. $f'(x) = (2x^2 + 3x + 8)[5] + (5x - 1)[4x + 3]$

3. $f'(x) = (x^2 + 5x + 3)^4[3(3x^2 - 2x + 6)^2(6x - 2)] + (3x^2 - 2x + 6)^3[4(x^2 + 5x + 3)^3(2x + 5)]$

5. $f'(x) = 5x\left[\dfrac{2x}{x^2 + 5}\right] + \ln(x^2 + 5)[5]$

7. $f'(x) = x^3[5e^{5x}] + e^{5x}[3x^2] + \dfrac{1}{x} + \dfrac{10}{7}x^{-2/7}$

9. First, since the numerator is a constant, it is easier to determine the derivative by rewriting as $3(x^5 + 7x - 3)^{-8}$. Therefore,

$$f'(x) = -24(x^5 + 7x - 3)^{-9}[5x^4 + 7] \text{ or } \frac{-24[5x^4 + 7]}{(x^5 + 7x - 3)^9}$$

11. $f'(x) = \dfrac{(5x^2 + 7x + 3)[3] - (3x + 2)[10x + 7]}{(5x^2 + 7x + 3)^2}$

13. $f'(x) = \dfrac{(5x^2 + 7x + 3)[12x^2 + 2] - (4x^3 + 2x + 1)[10x + 7]}{(5x^2 + 7x + 3)^2}$

Solutions Manual to Accompany Fundamentals of Calculus, First Edition. Carla C. Morris and Robert M. Stark.

15. $$f'(x) = \frac{(7x^3+5x+3)^4[11(5x^6+3x-4)^{10}(30x^5+3)]}{(7x^3+5x+3)^8} - \frac{(5x^6+3x-4)^{11}[4(7x^3+5x+3)^3(21x^2+5)]}{(7x^3+5x+3)^8}$$

17. Here, the power rule is applied first, and since the numerator is a constant, the expression in parenthesis is viewed as $30(x^7-5x^4+3)^{-3}$ to yield:

$$f'(x) = 100\left(\frac{30}{\left(x^7-5x^4+3\right)^3}\right)^{99}\left[\frac{-90\left(7x^6-20x^3\right)}{(x^7-5x^4+3)^4}\right]$$

19. $$f'(x) = \frac{(2x-5)\left[\frac{1}{x}\right] - \ln 7x[2]}{(2x-5)^2}$$

21. $f'(x) = 25(x^3+7x+\ln 8x)^{24}\left[3x^2+7+\frac{1}{x}\right]$

23. Here, the quotient rule is applied and the product rule for the derivative of the numerator:

$$f'(x) = \frac{(2x+7)^3[x^3(5e^{5x})+e^{5x}(3x^2))] - (x^3e^{5x})[3(2x+7)^2(2)]}{(2x+7)^6}$$

25. Here, the power rule is applied first and then the quotient rule is applied within the parenthesis to yield:

$$f'(x) = 4\left(\frac{\sqrt{3x-7}}{x^4+5x+3}\right)^3 \left[\frac{\left(x^4+5x+3\right)\left[\frac{1}{2}(3x-7)^{-1/2}(3)\right] - \sqrt{3x-7}[4x^3+5]}{(x^4+5x+3)^2}\right]$$

27. $$f'(x) = (3x+1)^4\left[10\left(\sqrt[5]{x^3}+2x\right)^9\left(\frac{3}{5}x^{-2/5}+2\right)\right] + \left(\sqrt[5]{x^3}+2x\right)^{10}[4(3x+1)^3(3)]$$

29. $f'(x) = (6x^3-\sqrt{x})^8[3x^2e^{x^3}] + e^{x^3}\left[8\left(6x^3-\sqrt{x}\right)^7\left(18x^2-\frac{1}{2}x^{-1/2}\right)\right]$

31. $f'(x) = \dfrac{(2x+1)^5[3e^{3x}] - e^{3x}[5(2x+1)^4(2)]}{(2x+1)^{10}}$

33. $f'(x) = \dfrac{(e^x+1)[4e^{4x}] - e^{4x}[e^x]}{(e^x+1)^2}$

35. First, $f'(x) = (5x-4)(3) + (3x+7)(5)$. Next, evaluating at $x = 1$ yields a slope of the tangent as $f'(1) = 1(3) + 10(5) = 53 = m$. The equation is $y - 10 = 53(x-1)$.

37. Here, $f'(x) = (x-1)^4[2] + (2x+3)[4(x-1)^3(1)]$. The slope is the derivative at $x = 2$ so, $f'(2) = (1)(2) + 7(4) = 30$. The equation is of the tangent line is $y - 7 = 30(x-2)$.

39. Here, $f'(x) = (4x-3)^{3/2}[1] + (x+1)[3/2(4x-3)^{1/2}(4)]$. The slope at x is $f'(3) = (27)(1) + 4(3/2)(3)(4) = 99$. The tangent line is $y - 108 = 99(x-3)$.

41. For a horizontal tangent, its slope is zero. Setting the first derivative to zero yields $0 = \dfrac{(2x-1)\lfloor 3x^2 \rfloor - x^3[2]}{(2x-1)^2}$. The fraction is zero when its numerator is zero, so $4x^3 - 3x^2 = 0$. This indicates a horizontal tangent at (0, 0) and $\left(\dfrac{3}{4}, \dfrac{27}{32}\right)$.

EXERCISES 5.2

1. Substituting $g(x)$ into $f(x)$ yields $f(g(x)) = (3x-1)^2 + 2(3x-1) + 4$.

3. Substituting $g(x)$ into $f(x)$ yields
$f(g(x)) = (9x^2+5x+7)^4 + 2(9x^2+5x+7)^3 + 5(9x^2+5x+7)$.

5. Letting $f(x) = x^4$ and $g(x) = 5x+1$ yields $f(g(x)) = h(x) = (5x+1)^4$.

7. Letting $f(x) = x^7 + \dfrac{2}{x^3}$ and $g(x) = 2x+1$ yields
$f(g(x)) = h(x) = (2x+1)^7 + \dfrac{2}{(2x+1)^3}$.

9. The function is a composite of x^{19} and $x^2 + 5x + 3$.
Using the chain rule, $f'(x) = 19(x^2+5x+3)^{18}(2x+5)$.

11. The function is a composite of x^5 and $5x^7 - 3x^4 + 3$.
Using the chain rule, $f'(x) = 5(5x^7 - 3x^4 + 3)^4(35x^6 - 12x^3)$.

13. The function is a composite of $x^{1/5}$ and $2x^4 + 3x^2 + 3$.
Using the chain rule, $f'(x) = \dfrac{1}{5}(2x^4 + 3x^2 + 3)^{-4/5}(8x^3 + 6x)$.

15. Here, $\dfrac{dy}{du} = \dfrac{5}{2}u^{3/2}$ and $\dfrac{du}{dx} = 8$. Multiplying,

⟶

$\frac{dy}{du} \cdot \frac{du}{dx} = \frac{5}{2}u^{3/2}(8) = 20u^{3/2}$. However, the answer is in terms of x, not u. Substitution yields $\frac{dy}{dx} = 20(8x-3)^{3/2}$.

17. Here, $\frac{dy}{du} = u[5(2u+3)^4(2)] + (2u+3)^5[1] = 3(2u+3)^4(4u+1)$ and $\frac{du}{dx} = 6x^2 + 7$. Multiplying, yields

$\frac{dy}{du} \cdot \frac{du}{dx} = 3(2u+3)^4(4u+1)(6x^2+7)$. However, the answer is in terms of x, not u. Substitution yields

$\frac{dy}{dx} = 3(4x^3+14x+5)^4(8x^3+28x+5)(6x^2+7)$.

19. Here, $\frac{dy}{du} = \frac{(u+1)[2]-(2u-1)[1]}{(u+1)^2} = \frac{3}{(u+1)^2}$ and $\frac{du}{dx} = 16x$.

Multiplying yields $\frac{dy}{du} \cdot \frac{du}{dx} = \frac{3}{(u+1)^2}(16x) = \frac{48x}{(u+1)^2}$. However,

the answer is in terms of x, not u. Substitution yields $\frac{dy}{dx} = \frac{48x}{(8x^2+4)^2}$.

EXERCISES 5.3

1. $8x + 18y\frac{dy}{dx} = 0$ so $\frac{dy}{dx} = \frac{-8x}{18y} = \frac{-4x}{9y}$

3. $6x^2 + 28y^3\frac{dy}{dx} = 3$ so $\frac{dy}{dx} = \frac{3-6x^2}{28y^3}$

5. $2x - 2 + 3y^2\frac{dy}{dx} = 3\frac{dy}{dx}$ so $\frac{2(x-1)}{3(1-y^2)} = \frac{dy}{dx}$

7. $\frac{-2}{x^3} - \frac{12}{y^5}\frac{dy}{dx} = 14$ so $\frac{dy}{dx} = \left(14 + \frac{2}{x^3}\right)\left(\frac{-y^5}{12}\right)$

9. $x^2\left[2y\frac{dy}{dx}\right] + y^2[2x] = 18$ so $\frac{dy}{dx} = \frac{18-2xy^2}{2x^2y} = \frac{9-xy^2}{x^2y}$

11. $x^3\left[5y^4\frac{dy}{dx}\right] + y^5[3x^2] + 2 = 12x$ so $\frac{dy}{dx} = \frac{12x-2-3x^2y^5}{5x^3y^4}$

13. $4x\left[3y^2\frac{dy}{dx}\right] + 4y^3 - x^2\left[\frac{dy}{dx}\right] - 2xy + 3x^2 = 9$ and rearranging terms

yields $(12xy^2 - x^2)\frac{dy}{dx} = -4y^3 + 2xy - 3x^2 + 9$. Therefore,

$\frac{dy}{dx} = \frac{-4y^3+2xy-3x^2+9}{12xy^2-x^2}$.

15. To solve, determine $\frac{dy}{dx}$ implicitly and then evaluate this derivative when $x = -3$ and $y = 5$. Therefore, $x\left[\frac{dy}{dx}\right] + y[1] = 0$ and $\frac{dy}{dx} = \frac{-y}{x}$. At $(-3,\ 5)$, $\frac{dy}{dx} = \frac{-5}{-3} = \frac{5}{3}$.

17. To solve, determine $\frac{dy}{dx}$ implicitly and then evaluate this derivative when $x = -1$ and $y = 1$. Therefore, $4y^3\frac{dy}{dx} + 2\frac{dy}{dx} - 9x^2 = 2$ so $\frac{dy}{dx} = \frac{2 + 9x^2}{4y^3 + 2}$. At $(-1,\ 1)$, $\frac{dy}{dx} = \frac{11}{6}$.

19. To solve, determine $\frac{dy}{dx}$ implicitly and then evaluate this derivative when $x = 5$ and $y = 2$. Therefore, $x\left[\frac{dy}{dx}\right] + y[1] + 3y^2\left[\frac{dy}{dx}\right] = 0$ and $\frac{dy}{dx} = \frac{-y}{x + 3y^2}$. At $(5, 2)$, $\frac{dy}{dx} = \frac{-2}{17}$.

21. Here, $x\left[4y^3\frac{dy}{dx}\right] + y^4 = 0$ and $\frac{dy}{dx} = \frac{-y^4}{4xy^3} = \frac{-y}{4x}$. At (3,2), $\frac{dy}{dx} = \frac{-2}{12} = \frac{-1}{6}$ so the equation of the tangent is $y - 2 = \frac{-1}{6}(x - 3)$.

23. $4x^3\frac{dx}{dt} + 4y^3\frac{dy}{dt} = 0$ and $\frac{dy}{dt} = \frac{-x^3}{y^3}\left(\frac{dx}{dt}\right)$

25. $2x\frac{dx}{dt} + 4x\frac{dy}{dt} + 4y\frac{dx}{dt} = 7\frac{dx}{dt} + 2y\frac{dy}{dt}$ or

$(2x + 4y - 7)\frac{dx}{dt} = (2y - 4x)\frac{dy}{dt}$ and therefore,

$\frac{dy}{dt} = \left(\frac{2x + 4y - 7}{2y - 4x}\right)\frac{dx}{dt}$.

27. The volume of a sphere is $V = \frac{4}{3}\pi r^3$. Half of the diameter yields a radius of 0.5 ft. Since air is being pumped into the balloon, its volume changes as $\frac{dV}{dt} = 3$. Differentiating, $\frac{dV}{dt} = 4\pi r^2\frac{dr}{dt}$. Solving, when $r = 0.5$ ft yields $\frac{dr}{dt} = \frac{3}{\pi}$ ft/min.

29. Given $I = 100/R$, and differentiating, $\frac{dI}{dR} = \frac{-100}{R^2}$. When $R = 5$, $\frac{dI}{dR} = -4$.

31. Using implicit differentiation

$$20x^{1/4}\left[\frac{3}{4}y^{-1/4}\frac{dy}{dx}\right]+20y^{3/4}\left[\frac{1}{4}x^{-3/4}\right]=0 \quad \text{so} \quad \frac{15x^{1/4}}{y^{1/4}}\frac{dy}{dx}=-\frac{5y^{3/4}}{x^{3/4}}$$

and $\frac{dy}{dx}=\frac{-y}{3x}$. Substituting $x=16$ and $y=81$ yields $\frac{dy}{dx}=\frac{-81}{48}$.

EXERCISES 5.4

1. $\Delta f(x)=(x+1)^3-x^3=x^3+3x^2+3x+1-x^3=3x^2+3x+1$

$$\begin{aligned}\Delta^2 f(x) &= f(x+1)-f(x)=3(x+1)^2+3(x+1)+1-(3x^2+3x+1)\\ &= 6x+6.\end{aligned}$$

An alternate method is to use

$$\begin{aligned}\Delta^2 f(x) &= f(x+2)-2f(x+1)+f(x)\\ &= (x+2)^3-2(x+1)^3+x^3=6x+6.\end{aligned}$$

3. First, one adds the functions as $f(x)+g(x)=x^2+3x+5+3^x$.

Next, differencing

$$\begin{aligned}&[(x+1)^2+3(x+1)+5]+3^{x+1}-(x^2+3x+5+3^x)\\ &=x^2+2x+1+3x+3+5+3^{x+1}-x^2-3x-5-3^x\\ &=2x+4+3^{x+1}-3^x\\ &=2x+4+2(3^x) \text{ (agreeing with Example 5.4.4)}\end{aligned}$$

First, subtract the functions as $f(x)-g(x)=x^2+3x+5-3^x$.

Next, differencing

$$\begin{aligned}&[(x+1)^2+3(x+1)+5]-3^{x+1}-(x^2+3x+5-3^x)\\ &=x^2+2x+1+3x+3+5-3^{x+1}-x^2-3x-5+3^x\\ &=2x+4-3^{x+1}+3^x\\ &=2x+4-2(3^x) \text{ (agreeing with Example 5.4.4)}\end{aligned}$$

First, multiply the functions as $f(x)g(x)=x^2(3^x)+3x(3^x)+5(3^x)$

Next, differencing

$$\begin{aligned}&(x+1)^2(3^{x+1})+3(x+1)(3^{x+1})+5(3^{x+1})-[x^2(3^x)+3x(3^x)+5(3^x)]\\ &=(x^2+2x+1+3x+3+5)(3^{x+1})-[(x^2+3x+5)(3^x)]\\ &=(x^2+5x+9)(3)(3^x)-[(x^2+3x+5)(3^x)]\\ &=(3x^2+15x+27-x^2-3x-5)(3^x)\\ &=(2x^2+12x+22)(3^x)\\ &=2(x^2+6x+11)(3^x) \text{ (agreeing with Example 5.4.4)}\end{aligned}$$

First, divide the functions as $\frac{f(x)}{g(x)}=\frac{x^2+3x+5}{3^x}$

⟶

Differencing as $\dfrac{(x+1)^2+3(x+1)+5}{3^{x+1}} - \dfrac{x^2+3x+5}{3^x}$

$$= \frac{(x+1)^2+3(x+1)+5}{3^{x+1}} - \frac{3(x^2+3x+5)}{3^{x+1}}$$

$$= \frac{x^2+2x+1+3x+3+5-3x^2-9x-15}{3^{x+1}}$$

$$= \frac{-2x^2-4x-6}{3^{x+1}} = \frac{-2(x^2+2x+3)}{3^{x+1}} \text{ (agreeing with Example 5.4.4)}$$

5. First, ${}_nP_r = \dfrac{n!}{(n-r)!} = n(n-1)\cdots(n-r+1)$

$$\Delta_n P_r = (n+1)(n)(n-1)\cdots((n+1)-r+1) - n(n-1)\cdots(n-r+1)$$
$$\Delta_n P_r = (n)(n-1)\cdots((n+1)-r+1)[(n+1)-(n-r+1]$$
$$= (n)(n-1)\cdots((n+1)-(r-1))[r]$$
$$\Delta_n P_r = (r)[{}_nP_{r-1}]$$

7. At a maximum, $f(x^*)$ is larger than any other value of the function, so $f(x^*) \geq f(x^*+1)$ and $f(x_*) \geq f(x^*-1)$. If $f(x^*) \geq f(x^*+1)$, subtract $f(x^*)$ from each side of the inequality $0 \geq f(x^*+1) - f(x^*) = f(x^*)$. Therefore, one condition for a maximum is that $\Delta f(x^*) \leq 0$. Another condition is $f(x^*) \geq f(x^*-1)$ and by subtracting $f(x^*-1)$ from each side of the inequality $f(x^*) - f(x^*-1) \geq 0$.
Since $f(x^*) - f(x^*-1) = \Delta f(x^*-1)$, another condition for a maximum is $\Delta f(x^*-1) \geq 0$. Combining both conditions yields the desired result $\Delta f(x^*) \leq 0 \leq \Delta f(x^*-1)$

9. When $a = 1$ and $b = 3$ then

$$\sum_{x=1}^{3} \Delta f(x)g(x) = \Delta[f(1)g(1)] + \Delta[f(2)g(2)] + [f(3)g(3)] = [f(2)g(2)$$
$$-f(1)g(1)] + [f(3)g(3) - f(2)g(2)] + [f(4)g(4) - f(3)g(3)]$$
$$= f(4)g(4) - f(1)g(1) = f(3+1)g(3+1) - f(1)g(1)$$
$$= f(b+1)g(b+1) - f(a)g(a)$$

11. a) One seeks $\sum_{x=1}^{n} x^2$. To evaluate the summation by antidifferences, seek an antidifference $F(x)$ such that $\Delta F(x) = x^2$. Since x^2 is a polynomial, it suggests as a trial, $F(x) = ax^3 + bx^2 + cx + d$.
The coefficients a, b, c, and d are determined so that $\Delta F(x) = x^2$. That is,

$$\Delta F(x) = [a(x+1)^3 + b(x+1)^2 + c(x+1) + d]$$
$$-[ax^3 + bx^2 + cx + d] = (3a)x^2 + (3a+2b)x + (a+b+c)$$

⟶

For $\Delta F(x)$ to equal x^2 requires $3a = 1$ and $(3a + 2b) = 0$, and $(a + b + c) = 0$.
Therefore, $a = \frac{1}{3}, \quad b = \frac{-1}{2}, \quad c = \frac{1}{6}$.
Now, one can write the following $\sum_{x=1}^{n} x^2 = \sum_{x=1}^{n} \Delta\left(\frac{1}{3}x^3 - \frac{1}{2}x^2 + \frac{1}{6}x\right)$, which yields

$$\left[\frac{1}{3}(n+1)^3 - \frac{1}{2}(n+1)^2 + \frac{1}{6}(n+1)\right] - \left[\frac{1}{3} - \frac{1}{2} + \frac{1}{6}\right]$$
$$= \frac{1}{3}n^3 + \frac{1}{2}n^2 + \frac{1}{6}n.$$

b) Using the result $a^x = \Delta\left(\frac{a^x}{a-1}\right)$ leads to $\sum_{x=1}^{n} a^x = \sum_{x=1}^{n} \Delta\left(\frac{a^x}{a-1}\right) = \frac{a^{n+1}}{a-1} - \frac{a^1}{a-1} = \frac{a^{n+1} - a}{a-1}$ and, after simplifying, $\frac{a(a^n - 1)}{a-1}$.

c) One seeks $\sum_{x=1}^{n} x^3$. To evaluate the summation by antidifferences, seek an antidifference $F(x)$ such that $\Delta F(x) = x^3$. Since x^3 is a polynomial, it suggests as a trial. $F(x) = ax^4 + bx^3 + cx^2 + dx + e$. The coefficients a, b, c, d, and e are determined so that $\Delta F(x) = x^3$. That is

$$\Delta F(x) = [a(x+1)^4 + b(x+1)^3 + c(x+1)^2 + d(x+1) + e]$$
$$- [ax^4 + bx^3 + cx^2 + dx + e]$$
$$= (4a)x^3 + (6a + 3b)x^2 + (4a + 3b + 2c)x + (a + b + c + d)$$

For $\Delta F(x)$ to equal x^3 requires
$4a = 1$, $(6a + 3b) = 0$, $(4a + 3b + 2c) = 0$, and
$(a + b + c + d) = 0$.
Therefore,

$$a = \frac{1}{4}, \quad b = \frac{-1}{2}, \quad c = \frac{1}{4}, \quad d = 0.$$

⟶

Now, one can write $\sum_{x=1}^{n} x^3 = \sum_{x=1}^{n} \Delta\left(\frac{1}{4}x^4 - \frac{1}{2}x^3 + \frac{1}{4}x^2\right)$, which yields

$$\left[\frac{1}{4}(n+1)^4 - \frac{1}{2}(n+1)^3 + \frac{1}{4}(n+1)^2\right] - \left[\frac{1}{4} - \frac{1}{2} + \frac{1}{4}\right] = \frac{1}{4}n^4 + \frac{1}{2}n^3 + \frac{1}{4}n^2.$$

SUPPLEMENTARY EXERCISES CHAPTER 5

1. $f'(x) = (5x^3 - 2x + 1)^4[2] + (2x+3)[4(5x^3 - 2x + 1)^3(15x^2 - 2)]$

3. $f'(x) = (2x^4 - 3x^2 + e^{3x})^2[3(2x+1)^2(2)] + (2x+1)^3[2(2x^4 - 3x^2 + e^{3x})(8x^3 - 6x + 3e^{3x})]$

5. $f'(x) = 10\left(\frac{5x+1}{3x-5}\right)^9 \left[\frac{(3x-5)[5] - (5x+1)[3]}{(3x-5)^2}\right]$

7. $f'(x) = \frac{(25x+3)^5 \left[8(2x+3+\ln 4x)^7 \left(2 + \frac{1}{x}\right)\right]}{(25x+3)^{10}} - \frac{(2x+3+\ln 4x)^8 \lfloor 5(25x+3)^4(25) \rfloor}{(25x+3)^{10}}$

9. $f'(x) = \frac{(4x^2 - 11x - 9)^6 \left[(x^3+5x+1)^3 \left(\frac{1}{x}\right) + (\ln 3x)3(x^3+5x+1)^2(3x^2+5)\right]}{(4x^2 - 11x - 9)^{12}} - \frac{(x^3+5x+1)^3 \ln 3x \lfloor 6(4x^2 - 11x - 9)^5(8x - 11) \rfloor}{(4x^2 - 11x - 9)^{12}}$

11. $f'(x) = (3x+1)^4 \left[\frac{15x^2 - 18x + 2}{5x^3 - 9x^2 + 2x + 1}\right] + \ln(5x^3 - 9x^2 + 2x + 1) [4(3x+1)^3(3)]$

13. When $x = 2$, by substitution $y = 11$. The derivative $f'(x) = (3x-5)^7[5] + (5x+1)[7(3x-5)^6(3)]$ evaluated when $x = 2$ yields 236. Therefore, one seeks a line through (2, 11) with slope 236. The tangent line is $y - 11 = 236(x - 2)$.

15. When $x = 0$, by substitution $y = 1$. The derivative $f'(x) = (x+1)^5[3e^{3x}] + (e^{3x})[5(x+1)^4]$ evaluated when $x = 0$

yields 8. Therefore, one seeks a line through (0, 1) with slope 8. The tangent is $y - 1 = 8(x - 0)$ or $y = 8x + 1$.

17. $\frac{dy}{du} = 14u^6$ and $\frac{du}{dx} = 15x^4 - 18x + 5$ so

$\frac{dy}{du} \cdot \frac{du}{dx} = 14u^6(15x^4 - 18x + 5)$. Writing, in terms of x, yields $\frac{dy}{dx} = 14(3x^5 - 9x^2 + 5x + 10)^6(15x^4 - 18x + 5)$.

19. a) $f(g(x)) = f(x^7 + 2x^5 + x^2 + 1) = (x^7 + 2x^5 + x^2 + 1)^{10}$

b) The derivative of x^{10} is $10x^9$ so the chain rule begins with $10(x^7 + 2x^5 + x^2 + 1)^9$ and if multiplied by $(7x^6 + 10x^4 + 2x)$ to yield $10(x^7 + 2x^5 + x^2 + 1)^9(7x^6 + 10x^4 + 2x)$

c) The derivative of $f(g(x))$ is $10(x^7 + 2x^5 + x^2 + 1)^9(7x^6 + 10x^4 + 2x)$, which agrees with part b.

21. $3x^4\left[3y^2\frac{dy}{dx}\right] + y^3[12x^3] = 10y\left[\frac{dy}{dx}\right] + 12x$ and rearranging terms, $(9x^4y^2 - 10y)\frac{dy}{dx} = 12x - 12x^3y^3$. Therefore, $\frac{dy}{dx} = \frac{12x - 12x^3y^3}{9x^4y^2 - 10y}$.

23. Here, $p\left(\frac{dv}{dt}\right) + v\left(\frac{dp}{dt}\right) = 0$ and rearranging yields $v\left(\frac{dp}{dt}\right) = -p\left(\frac{dv}{dt}\right)$ or that $\left(\frac{dp}{dt}\right) = -\frac{p}{v}\left(\frac{dv}{dt}\right)$. Given $v = 75$, $p = 30$ and $\frac{dv}{dt} = 5$, substituting yields $\frac{dp}{dt} = -2$. The pressure is decreasing at 2lbs/in.^{2}min.

25.
$$\begin{aligned}\Delta f(x) &= f(x+1) - f(x) = (x+1)^4 - x^4 = x^4 + 4x^3 + 6x^2 + 4x + 1 - x^4\\ &= 4x^3 + 6x^2 + 4x + 1\\ \Delta^2 f(x) &= \Delta(\Delta f(x))\\ &= 4(x+1)^3 + 6(x+1)^2 + 4(x+1) + 1 - [4x^3 + 6x^2 + 4x + 1]\\ &= 12x^2 + 24x + 14\end{aligned}$$

CHAPTER 6

INTEGRAL CALCULUS

EXERCISES 6.1

1. $\int 7dx = 7x + C$

3. $\int 5xdx = \frac{5x^2}{2} + C$

5. $\int 3x^{-5}dx = -\frac{3x^{-4}}{4} + C$

7. $\int 2tdt = t^2 + C$

9. First, rewrite as $\int x^{1/2}dx$ to yield $\int x^{1/2}dx = \frac{2}{3}x^{3/2} + C$

11. First, rewrite as $\int x^{2/3}dx$ to yield $\int x^{2/3}dx = \frac{3}{5}x^{5/3} + C$

13. First, rewrite as $\int \frac{1}{2}x^{-3}dx$ to yield $\int \frac{1}{2}x^{-3}dx = -\frac{1}{4}x^{-2} + C$

15. Integrating term by term yields

$$\int \left(x^{3/5} + x^{-2/3}\right)dx = \frac{5}{8}x^{8/5} + 3x^{1/3} + C$$

Solutions Manual to Accompany Fundamentals of Calculus, First Edition. Carla C. Morris and Robert M. Stark.

17. $\int e^{7t}dt = \frac{e^{7t}}{7} + C$

19. Integrating term by term yields
$\int (4x^3 + 3x^2 + 2x + 9)dx = x^4 + x^3 + x^2 + 9x + C$

21. Integrating term by term yields $\int (e^{4x} + 1)dx = \frac{e^{4x}}{4} + x + C$

23. Integrating term by term yields $\int \left(4 - \frac{2}{x}\right)dx = 4x - 2\ln|x| + C$

25. Rewrite $\int\left(\frac{3}{\sqrt{t}} - 2\sqrt{t}\right)dt$ as $\int (3t^{-1/2} - 2t^{1/2})dt$. Integration yields
$\int (3t^{-1/2} - 2t^{1/2})dt = 6t^{1/2} - \frac{4}{3}t^{3/2} + C$

27. First, integration yields $f(x) = x^3 + C$. Next, the initial condition indicates that $6 = (-1)^3 + C$ and that $C = 7$, so $f(x) = x^3 + 7$ is the desired function.

29. First, integration yields $f(x) = x^3 - x^2 + 4x + C$. Next, the initial condition indicates that $10 = (2)^3 - (2)^2 + 4(2) + C$ and that $C = -2$, so $f(x) = x^3 - x^2 + 4x - 2$ is the desired function.

31. Differentiating a) yields $\frac{5}{2}[2xe^{x^2}] = 5xe^{x^2}$ and b) yields $5x[2xe^{x^2}] + e^{x^2}[5]$, therefore, part a) indicates that
$\int 5xe^{x^2}dx = \frac{5}{2}e^{x^2} + C$

EXERCISES 6.2

1. The subinterval width, Δx, is $\frac{b-a}{n}$, where a and b are endpoints. Here, $\Delta x = \frac{4-0}{4} = 1$. Divide the interval 0 to 4 into four subintervals: [0, 1], [1, 2], [2, 3], and [3, 4]. The four left endpoints are 0, 1, 2, and 3.

3. The subinterval width, Δx, is $\frac{b-a}{n}$, where a and b are endpoints. Here, $\Delta x = \frac{3-0}{4} = \frac{3}{4}$. Divide the interval 0 to 3 into four subintervals: [0, 0.75], [0.75, 1.5], [1.5, 2.25], and [2.25, 3]. The four left endpoints are 0, 0.75, 1.5, and 2.25.

5. The subinterval width, Δx, is $\frac{b-a}{n}$, where a and b are endpoints. Here, $\Delta x = \frac{15-1}{7} = 2$. Divide the interval 1 to 15 into seven subintervals: [1, 3], [3, 5], [5, 7], [7, 9], [9, 11], [11, 13], and [13, 15]. The seven right endpoints are 3, 5, 7, 9, 11, 13, and 15.

7. The subinterval width, Δx, is $\frac{b-a}{n}$, where a and b are endpoints. Here, $\Delta x = \frac{27-3}{6} = 4$. Divide the interval 3 to 27 into six subintervals: [3, 7], [7, 11], [11, 15], [15, 19], [19, 23], and [23, 27]. The six right endpoints are 7, 11, 15, 19, 23, and 27.

9. The subinterval width, Δx, is $\frac{b-a}{n}$, where a and b are endpoints. Here, $\Delta x = \frac{4-0}{8} = \frac{1}{2}$. Divide the interval 0 to 4 into eight subintervals: [0, 0.5], [0.5, 1.0], [10, 1.5], [1.5, 2.0], [2.0, 2.5], [2.5, 3.0], [3.0, 3.5], and [3.5, 4.0]. The eight midpoints are 0.25, 0.75, 1.25, 1.75, 2.25, 2.75, 3.25, and 3.75.

11. The subinterval width, Δx, is $\frac{b-a}{n}$, where a and b are endpoints. Here, $\Delta x = \frac{21-5}{4} = 4$. Divide the interval 5 to 21 into four subintervals: [5, 9], [9, 13], [13, 17], and [17, 21]. The four midpoints are 7, 11, 15, and 19.

13. The subinterval width, Δx, is $\frac{b-a}{n}$, where a and b are endpoints. Here, $\Delta x = \frac{91-1}{4} = 2$. Divide the interval 1 to 9 into four subintervals: [1, 3], [3, 5], [5, 7], and [7, 9]. The four right endpoints are 3, 5, 7, and 9. Next, use $\frac{b-a}{n}\sum_{i=1}^{4} f(x_i)$ to yield
$2[f(3)+f(5)+f(7)+f(9)] = 2[10+26+50+82] = 336.$

15. The subinterval width, Δx, is $\frac{b-a}{n}$, where a and b are endpoints. Here, $\Delta x = \frac{20-0}{5} = 4$. Divide the interval 0 to 20 into five subintervals: [0, 4], [4, 8], [8, 12], [12, 16], and [16, 20]. The five right endpoints are 4, 8, 12, 16, and 20. Next, use

⟶

$\frac{b-a}{n}\sum_{i=1}^{5} f(x_i)$ to yield $4[f(4)+f(8)+f(12)+f(16)+f(20)]$
$= 4[64 + 512 + 1728 + 4096 + 8000] = 57,600.$

17. The subinterval width, Δx, is $\frac{b-a}{n}$, where a and b are endpoints. Here, $\Delta x = \frac{23-3}{4} = 5$. Divide the interval 3 to 23 into four subintervals: [3, 8], [8, 13], [13, 18], and [18, 23]. The four left endpoints are 3, 8, 13, and 18. Next, use $\frac{b-a}{n}\sum_{i=1}^{4} f(x_i)$ to yield $5[f(3)+f(8)+f(13)+f(18)] = 5[19 + 44 + 69 + 94] = 1130.$

19. The subinterval width, Δx, is $\frac{b-a}{n}$, where a and b are endpoints. Here, $\Delta x = \frac{9-1}{4} = 2$. Divide the interval 1 to 9 into four subintervals: [1, 3], [3, 5], [5, 7], and [7, 9]. The four left endpoints are 1, 3, 5, and 7. Next, use $\frac{b-a}{n}\sum_{i=1}^{4} f(x_i)$ to yield
$2[f(1)+f(3)+f(5)+f(7)] = \frac{352}{105} \approx 3.35.$

21. The subinterval width, Δx, is $\frac{b-a}{n}$, where a and b are endpoints. Here, $\Delta x = \frac{9-1}{4} = 2$. Divide the interval 1 to 9 into four subintervals: [1, 3], [3, 5], [5, 7], and [7, 9]. The four midpoints are 2, 4, 6, and 8. Next, use $\frac{b-a}{n}\sum_{i=1}^{4} f(x_i)$ to yield $2[f(2)+f(4)+f(6)+f(8)] = 248.$

23. The subinterval width, Δx, is $\frac{b-a}{n}$, where a and b are endpoints. Here, $\Delta x = \frac{3-1}{4} = \frac{1}{2}$. Divide the interval 1 to 3 into four subintervals: [1, 1.5], [1.5, 2], [2, 2.5], and [2.5, 3]. The four midpoints are 1.25, 1.75, 2.25, and 2.75. Next, use
$\frac{b-a}{n}\sum_{i=1}^{4} f(x_i)$ to yield $\frac{1}{2}[f(1.25)+f(1.75)+f(2.25)+f(2.75)]$
$= \frac{1}{2}[e^5 + e^7 + e^9 + e^{11}] \approx 34,611.14.$

25. The subinterval width, Δx, is $\frac{b-a}{n}$, where a and b are endpoints. Here, $\Delta x = \frac{7-1}{6} = 1$. Divide the interval 1 to 7 into six subintervals [1, 2], [2, 3], [3, 4], [4, 5], [5, 6], and [6, 7]. The six right endpoints are 2, 3, 4, 5, 6, and 7. Next, use $\frac{b-a}{n}\sum_{i=1}^{6} f(x_i)$ to yield $1[f(2)+f(3)+f(4)+f(5)+f(6)+f(7)]$. Using the graph yields $1[6+10+7+4+7+10] = 44$.

27. The subinterval width, Δx, is $\frac{b-a}{n}$, where a and b are endpoints. Here, $\Delta x = \frac{10-2}{8} = 1$. Divide the interval 2 to 10 into eight subintervals: [2, 3], [3, 4], [4, 5], [5, 6], [6, 7], [7, 8], [8, 9], and [9, 10]. The eight left endpoints are 2, 3, 4, 5, 6, 7, 8, and 9. Next, use $\frac{b-a}{n}\sum_{i=1}^{8} f(x_i)$ to yield $1[f(2)+f(3)+f(4)+f(5)+f(6)+f(7)+f(8)+f(9)]$ and using the graph yields $1[6+10+7+4+7+10+14+12] = 70$.

29. The subinterval width, Δx, is $\frac{b-a}{n}$, where a and b are endpoints. Here, $\Delta x = \frac{9-1}{4} = 2$. Divide the interval 1 to 9 into four subintervals: [1, 3], [3, 5], [5, 7], and [7, 9]. The four midpoints are 2, 4, 6, and 8. Next, use $\frac{b-a}{n}\sum_{i=1}^{4} f(x_i)$ to yield $2[f(2)+f(4)+f(6)+f(8)]$ and using the graph yields $2[6+7+7+14] = 68$.

31. a) Using a rectangle and a right triangle the area, geometrically, is $(1 \times 4) + (1/2)(4 \times 16) = 36$ square units.

b) Using a Riemann sum, $\frac{b-a}{n}$ is unity and the four right endpoints are 1, 2, 3, and 4. The approximate area is $1[f(1)+f(2)+f(3)+f(4)] = 1[5+9+13+17] = 44$. This overestimates the actual area since $4x + 1$ is increasing on the interval [0, 4].

c) $\int_0^4 (4x+1)dx = (2x^2+x)|_0^4 = 36$

EXERCISES 6.3

1. $\int_4^9 dx = x|_4^9 = 9 - 4 = 5$

3. $\int_1^4 3\,dx = 3x|_1^4 = 12 - 3 = 9$

5. $\int_2^5 (2x+3)dx = \left(x^2+3x\right)\Big|_2^5 = 40 - 10 = 30$

7. $\int_1^2 (4x^3+2x+5)dx = x^4 + x^2 + 5x|_1^2$

$= (16+4+10) - (1+1+5) = 23$

9. $\int_0^{15} (e^{3p})dp = \left(\dfrac{e^{3p}}{3}\right)\Bigg|_0^{15} = \dfrac{1}{3}(e^{45} - 1)$

11. $\int_3^6 \dfrac{1}{t}\,dt = \ln t|_3^6 = \ln 6 - \ln 3 = \ln 2$

13. $\int_1^3 \left(\dfrac{4}{t^2}\right) dt = \left(-\dfrac{4}{t}\right)\Bigg|_1^3 = -\dfrac{4}{3} + 4 = \dfrac{8}{3}$

15. $\int_{-1}^4 (8x^3+5x+2)dx = \left(2x^4 + \dfrac{5x^2}{2} + 2x\right)\Bigg|_{-1}^4$

$= (512+40+8) - \left(2 + \dfrac{5}{2} - 2\right) = \dfrac{1115}{2}$

17. $\int_1^9 \sqrt{t}\,dt = \left(\dfrac{2}{3}t^{3/2}\right)\Bigg|_1^9 = 18 - \dfrac{2}{3} = \dfrac{52}{3}$

19. $\int_1^8 \sqrt[3]{x^2}\,dx = \left(\dfrac{3}{5}x^{5/3}\right)\Bigg|_1^8 = \dfrac{96}{5} - \dfrac{3}{5} = \dfrac{93}{5}$

21. $\int_1^8 \left(x^3 + 3x^2 + \sqrt[3]{x}\right) dx = \left(\dfrac{x^4}{4} + x^3 + \dfrac{3}{4}x^{4/3}\right)\Bigg|_1^8$

$= (1024+512+12) - \left(\dfrac{1}{4} + 1 + \dfrac{3}{4}\right) = 1546$

23. $\int_0^3 \left(4 - \dfrac{x^2}{4}\right)dx = \left(4x - \dfrac{x^3}{12}\right)\Bigg|_0^3 = \left(12 - \dfrac{27}{12}\right) - (0-0) = \dfrac{39}{4}$

25. In the graph (below) the rectangular area between $x = 2$ and $x = 7$ with a height of 6 is $6(5) = 30$.

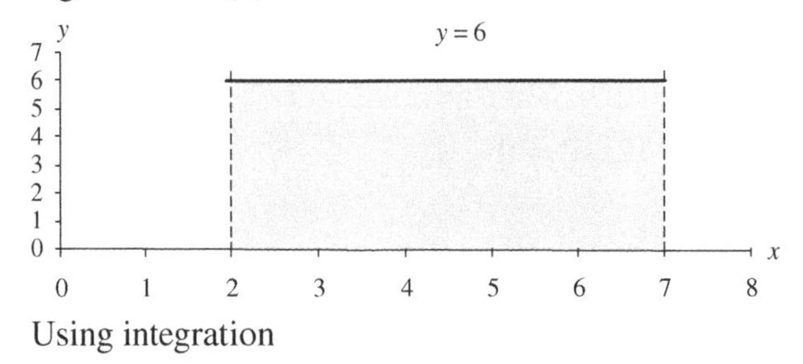

Using integration

$$\int_2^7 6\,dx = (6x)|_2^7 = 42 - 12 = 30.$$

27. In the graph (below) the area between $x = 1$ and $x = 3$ is

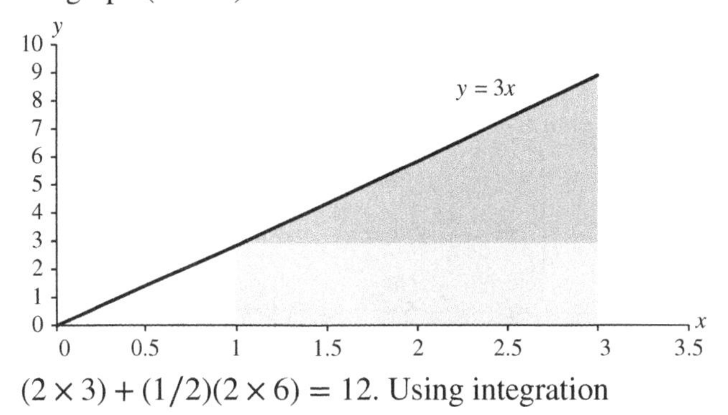

$(2 \times 3) + (1/2)(2 \times 6) = 12$. Using integration

$$\int_1^3 3x\ dx = \left(\frac{3x^2}{2}\right)\Bigg|_1^3 = \frac{27}{2} - \frac{3}{2} = 12.$$

EXERCISES 6.4

1. Setting $8 - x^2 = -2x + 5$ yields $x^2 - 2x - 3 = 0$. This factors as $(x - 3)(x + 1) = 0$. The limits of integration are -1 and 3. The area bounded by the curves is

$\int_{-1}^{3} \left[\left(8 - x^2\right) - (-2x + 5)\right] dx$. Integration yields

$$\int_{-1}^{3} \left[-x^2 + 2x + 3\right] dx = \left(\frac{-x^3}{3} + x^2 + 3x\right)\Bigg|_{-1}^{3} = \frac{32}{3}.$$

3. Setting $x^2 = \sqrt{x}$ yields $x^2 - \sqrt{x} = 0$. Solve this equation for the limits of integration 0 and 1. The area bounded by the curves is $\int_0^1 (\sqrt{x} - x^2)dx$. Integration yields

$$\int_0^1 \left(\sqrt{x} - x^2\right) dx = \left(\frac{2x^{3/2}}{3} - \frac{x^3}{3}\right)\Bigg|_0^1 = \frac{1}{3}.$$

5. Setting $x^2 + 3 = 4x + 3$ yields $x^2 - 4x = 0$. This factors as $(x)(x - 4) = 0$. The limits of integration are 0 and 4.

The area bounded by the curves is $\int_0^4 \left[(4x + 3) - (x^2 + 3)\right] dx$.

Integration yields $\int_0^4 \left(4x - x^2\right) dx = \left(2x^2 - \frac{-x^3}{3}\right)\Bigg|_0^4 = \frac{32}{3}$.

7. Setting $x^2 - 3x + 1 = -x + 4$ yields $x^2 - 2x - 3 = 0$. This factors as $(x - 3)(x + 1) = 0$.

The limits of integration are -1 and 3. The area bounded by the curves is $\int_{-1}^3 \left[(-x + 4) - (x^2 - 3x + 1)\right] dx$.

Integration yields

$$\int_{-1}^3 \left[-x^2 + 2x + 3)\right] dx = \left(\frac{-x^3}{3} + x^2 + 3x\right)\Bigg|_{-1}^3 = \frac{32}{3}.$$

9. Setting $x^2 - 4x + 3 = -x^2 + x + 3$ yields $2x^2 - 5x = 0$. This factors as $(x)(2x - 5) = 0$. The limits of integration are 0 and 5/2.

The area bounded by the curves is

$$\int_0^{5/2} \left[\left(-x^2 + x + 3\right) - \left(x^2 - 4x + 3\right)\right] dx.$$

Integration yields

$$\int_0^{5/2} \left[-2x^2 + 5x\right] dx = \left(\frac{-2x^3}{3} + \frac{5x^2}{2}\right)\Bigg|_0^{5/2} = \frac{125}{24}.$$

11. Since the curve $y = x^2$ is above the x-axis on the interval $(-2, 3)$, one integrates $\int_{-2}^3 (x^2)dx = \left(\frac{x^3}{3}\right)\Bigg|_{-2}^3 = \frac{35}{3}$.

13. Setting $x^3 - x^2 + 1 = 2x + 1$ yields $x^3 - x^2 - 2x = 0$. This factors as $(x)(x - 2)(x + 1) = 0$. The limits of integration are -1, 0, and 2.

The area bounded by the curves is

$$\int_{-1}^{0} \left[\left(x^3 - x^2 + 1\right) - (2x + 1)\right] dx + \int_{0}^{2} \left[(2x + 1) - (x^3 - x^2 + 1)\right] dx.$$

Integration yields

$$\int_{-1}^{0} [(x^3 - x^2 - 2x)]dx + \int_{0}^{2} [(-x^3 + x^2 + 2x)]dx$$

$$= \left(\frac{x^4}{4} - \frac{x^3}{3} - x^2\right)\Bigg|_{-1}^{0} + \left(\frac{-x^4}{4} + \frac{x^3}{3} + x^2\right)\Bigg|_{0}^{2} = \frac{37}{12}$$

15. The average value is

$$\frac{1}{4-1}\int_{1}^{4} (4x + 3)\, dx = \frac{1}{3}\left(2x^2 + 3x\right)\Big|_{1}^{4} = 13.$$

17. The average value is

$$\frac{1}{6-2}\int_{2}^{6} \left(\frac{1}{x}\right) dx = \frac{1}{4}\,(\ln x)|_{2}^{6} = \frac{1}{4}\,(\ln 6 - \ln 2) = \frac{1}{4}\ln 3.$$

19. The average value is $\frac{1}{8-0}\int_{0}^{8} \left(x^{1/3}\right) dx = \frac{3}{32}x^{4/3}|_{0}^{8} = \frac{3}{2}.$

21. The average value is

$$\frac{1}{20-6}\int_{6}^{20} \left(\frac{1}{2}x^2 + x + 100\right) dx = \frac{1}{14}\left(\frac{x^3}{6} + \frac{x^2}{2} + 100x\right)\Bigg|_{6}^{20} \approx 205.67.$$

23. One seeks

$$\int_{0}^{5} \left[\left(-x^2 + 34\right) - (9)\right] dx = \int_{0}^{5} \left(25 - x^2\right) dx = \left(25x - \frac{x^3}{3}\right)\Bigg|_{0}^{5} = \frac{250}{3}.$$

SUPPLEMENTARY EXERCISES CHAPTER 6

1. Integration yields $\int 4t^3 dt = t^4 + C$. The derivative of $t^4 + C$ is $4t^3$, which is in agreement with the integrand.

3. Integration yields $\int (e^{3x} + 6x + 5)dx = \frac{e^{3x}}{3} + 3x^2 + 5x + C.$

5. The width of the subintervals is $\Delta x = \frac{21-3}{6} = 3.$

The right endpoints are 6, 9, 12, 15, 18, and 21.

7. The width of the subintervals is $\Delta x = \frac{11-3}{4} = 2.$The left endpoints are 3, 5, 7, and 9.

The fourth Riemann Sum is

$2[f(3) + f(5) + f(7) + f(9)] = 2[9 + 25 + 49 + 81] = 328.$

9. Integration yields

$$\int_1^4 (5 + 2e^{3x})dx = \left(5x + \frac{2e^{3x}}{3}\right)\Bigg|_1^4 = 15 + \frac{2}{3}(e^{12} - e^3)$$

11. Integration yields $\int_1^4 x^{3/2}\ dx = \left(\frac{2}{5}x^{5/2}\right)\Big|_1^4 = \frac{62}{5}$

13. Setting $x + 1 = x^3 + 1$ yields $x^3 - x = 0$. This factors as $(x)(x - 1)(x + 1) = 0$.

The limits of integration are -1, 0, and 1. The area bounded by the curves is

$$\int_{-1}^0 [(x^3 + 1) - (x + 1)]dx + \int_0^1 [(x + 1) - (x^3 + 1)]dx.$$

Integration yields $\int_{-1}^0 [(x^3 - x)]dx + \int_0^1 [(-x^3 + x)]dx$

$$= \left(\frac{x^4}{4} - \frac{x^2}{2}\right)\Bigg|_{-1}^0 + \left(\frac{-x^4}{4} + \frac{x^2}{2}\right)\Bigg|_0^1 = \frac{1}{2}.$$

15. Setting $x^2 - 7 = -x^2 + 2x + 5$ yields $2x^2 - 2x - 12 = 0$. This factors as $2(x - 3)(x + 2) = 0$ and the limits of integration are -2 and 3. The area bounded by the curves is

$$\int_{-2}^3 [-x^2 + 2x + 5) - (x^2 - 7)]dx = \int_{-2}^3 (-2x^2 + 2x + 12)dx$$

Integration yields

$$\int_{-2}^3 (-2x^2 + 2x + 12)dx = \left(\frac{-2x^3}{3} + x^2 + 12x\right)\Bigg|_{-2}^3 = \frac{125}{3}.$$

17. The average value is

$$\frac{1}{2 - 0}\int_0^2 (4 - x^2)dx = \frac{1}{2}\left(4x - \frac{x^3}{3}\right)\Bigg|_0^2 = \frac{8}{3}.$$

CHAPTER 7

INTEGRATION TECHNIQUES

EXERCISES 7.1

1. Let $u = 5x + 3$, $du = 5dx$, and $\frac{1}{5}du = dx$. Therefore,
$\int (5x+3)^{-3/4}dx = \int u^{-3/4}\left(\frac{1}{5}du\right)$. Rewriting as $\frac{1}{5}\int u^{-3/4}du$
yields $\frac{4}{5}u^{1/4} + C$. The expression is in terms of x so

$$\int (5x+3)^{-3/4}dx = \frac{4}{5}(5x+3)^{1/4} + C.$$

3. Let $u = x^2 + 5$, $du = 2xdx$, and $\frac{1}{2}du = xdx$. Therefore,
$\int x(x^2+5)^4dx = \int u^4\left(\frac{1}{2}du\right)$

Rewriting as $\frac{1}{2}\int u^4du$ yields $\frac{1}{10}u^5 + C$. The expression is in terms of x so $\int x(x^2+5)^4dx = \frac{1}{10}(x^2+5)^5 + C$.

Solutions Manual to Accompany Fundamentals of Calculus, First Edition. Carla C. Morris and Robert M. Stark.

5. Let $u = x^4 + x^2 + 11$, $du = (4x^3 + 2x)dx$, and $2du = (8x^3 + 4x)dx$. Therefore, $\int (8x^3 + 4x)(x^4 + x^2 + 11)^9\, dx = \int u^9 2du$

Rewriting as $2\int u^9 du$ yields $\frac{1}{5}u^{10} + C$. The expression is in terms of x so $\int (8x^3 + 4x)(x^4 + x^2 + 11)^9\, dx = \frac{1}{5}(x^4 + x^2 + 11)^{10} + C$.

7. Let $u = x^2$, $du = (2x)dx$, and $4du = 8xdx$. Therefore,
$\int 8xe^{x^2} dx = \int e^u(4du)$

Rewriting as $4\int e^u du$ yields $4e^u + C$. The expression is in terms of x so $\int 8xe^{x^2} dx = 4e^{x^2} + C$.

9. Let $u = x^2 + 3x + 5$ and $du = (2x + 3)dx$. Therefore,
$\int \frac{2x+3}{x^2+3x+5} = \int \frac{1}{u}du = \ln|u| + C$. The expression is in terms of x so $\int \frac{2x+3}{x^2+3x+5} = \ln|x^2 + 3x + 5| + C$.

11. Let $u = x^3 - 3x^2 + 1$, $du = (3x^2 - 6x)dx$, and $\frac{1}{3}du = (x^2 - 2x)dx$. Therefore, $\int \frac{x^2 - 2x}{x^3 - 3x^2 + 1} dx = \int \frac{1}{u}\left(\frac{1}{3}du\right)$.

Rewriting as $\frac{1}{3}\int \frac{1}{u}du$ yields $\frac{1}{3}\ln|u| + C$. The expression is in terms of x so $\int \frac{x^2 - 2x}{x^3 - 3x^2 + 1}\, dx = \frac{1}{3}\ln|x^3 - 3x^2 + 1| + C$.

13. Let $u = x^5 + 1$, $du = (5x^4)dx$, and $\frac{1}{5}du = (x^4)dx$. Therefore,
$\int \frac{x^4}{x^5+1}dx = \int \frac{1}{u}\left(\frac{1}{5}du\right)$

Rewriting as $\frac{1}{5}\int \frac{1}{u}du$ yields $\frac{1}{5}\ln|u| + C$. The expression is in terms of x so $\int \frac{x^4}{x^5+1}\, dx = \frac{1}{5}\ln|x^5 + 1| + C$.

15. Let $u = x^2 + 2$, $du = 2xdx$, and $\frac{1}{2}du = xdx$. Therefore,

$\int \left(\frac{x}{\sqrt{x^2+2}}\right)dx = \int u^{-1/2}\left(\frac{1}{2}du\right)$.

$\longrightarrow$

Rewriting as $\frac{1}{2}\int u^{-1/2}\,du$ yields $u^{1/2} + C$. The expression is in terms of x so $\int \left(\frac{x}{\sqrt{x^2+2}}\right) dx = \sqrt{x^2+2} + C$.

17. Let $u = x^2 + 9$, $du = 2xdx$, and $3du = 6xdx$. Therefore, $\int 6x\sqrt{x^2+9}dx = \int u^{1/2}(3du)$. Rewriting as $3\int u^{1/2}\,du$ yields $2u^{3/2} + C$. The expression is in terms of x so

$$\int 6x\sqrt{x^2+9}\ dx = 2(x^2+9)^{3/2} + C.$$

19. Let $u = 3x^4 + 5x^2 + 8$, $du = (12x^3 + 10x)dx$, and $\frac{1}{2}du = (6x^3 + 5x)dx$. Therefore, $\int (6x^3 + 5x)(3x^4 + 5x^2 + 8)^{10}dx = \int u^{10}\left(\frac{1}{2}du\right)$

Rewriting as $\frac{1}{2}\int u^{10}du$ yields $\frac{1}{22}u^{11} + C$. The expression is in terms of x so

$$\int (6x^3 + 5x)(3x^4 + 5x^2 + 8)^{10}dx = \frac{1}{22}(3x^4 + 5x^2 + 8)^{11} + C.$$

EXERCISES 7.2

1. Using integration by parts, let $f(x) = x, f'(x) = 1\,dx, g'(x) = e^{9x}\,dx$, and $g(x) = \frac{e^{9x}}{9}$. Therefore,

$$\int xe^{9x}dx = x\left(\frac{e^{9x}}{9}\right) - \int \left(\frac{e^{9x}}{9}\right)(1dx) = \frac{xe^{9x}}{9} - \frac{e^{9x}}{81} + C.$$

3. Using integration by parts, let $f(x) = x$, $f'(x) = dx$, $g'(x) = e^{-x}dx$, and $g(x) = -e^{-x}$. Therefore,

$$\int xe^{-x}dx = -xe^{-x} + \int e^{-x}dx = -xe^{-x} - e^{-x} + C.$$

5. Using integration by parts for the first term, let

$f(x) = x$, $f'(x) = dx$, $g'(x) = e^{7x}dx$, and $g(x) = \frac{e^{7x}}{7}$. Therefore,

$$\int (xe^{7x} + 4x + 3)dx = \int xe^{7x}dx + \int (4x + 3)dx$$

$$= \frac{xe^{7x}}{7} - \int \left(\frac{e^{7x}}{7}\right)dx + 2x^2 + 3x + C$$

$$= \frac{xe^{7x}}{7} - \frac{e^{7x}}{49} + 2x^2 + 3x + C.$$

7. Using integration by parts, let

$f(x) = \ln 5x, f'(x) = \frac{1}{x}dx$ $g'(x) = x^3dx$, and $g(x) = \frac{x^4}{4}$. Therefore,

$$\int (x^3 \ln 5x)dx = \frac{x^4}{4} \ln 5x - \int \frac{x^3}{4}dx = \frac{x^4}{4} \ln 5x - \frac{x^4}{16} + C.$$

9. Using integration by parts, let

$f(x) = \ln 9x, f'(x) = \frac{1}{x}\,dx\,g'(x) = 6x^5\,dx$, and $g(x) = x^6$. Therefore,

$$\int (6x^5 \ln 9x)\ dx = x^6 \ln 9x - \int x^5dx = x^6 \ln 9x - \frac{x^6}{6} + C.$$

11. Using integration by parts for the first term, let

$f(x) = \ln 3x, f'(x) = \frac{1}{x}\,dx\,g'(x) = x^8$, and $g(x) = \frac{x^9}{9}$. Therefore,

$$\int (x^8 \ln 3x + e^{2x} + 6)\ dx = \int x^8 \ln 3xdx + \int (e^{2x} + 6)\,dx$$

$$= \frac{x^9}{9} \ln 3x - \left(\int \frac{x^8}{9}dx\right) + \frac{e^{2x}}{2} + 6x + C$$

$$= \frac{x^9}{9} \ln 3x - \frac{x^9}{81} + \frac{e^{2x}}{2} + 6x + C.$$

13. Using integration by parts
$f(x) = 5x$, $f'(x) = 5dx$, $g'(x) = (x + 2)^{-1/2}dx$, and $g(x) = 2(x + 1)^{1/2}$. Therefore,

$$\int \frac{5x}{\sqrt{x+2}}dx = \int 5x(x+2)^{-1/2}dx$$

$$= 10x(x+1)^{1/2} - \int 10(x+2)^{1/2}dx$$

$$= 10x(x+2)^{1/2} - \frac{20}{3}(x+2)^{3/2} + C.$$

15. Using integration by parts $f(x) = x$, $f'(x) = dx$, $g'(x) = (x + 4)^{-2}dx$, and $g(x) = -(x + 4)^{-1}$. Therefore,

$$\int x(x+4)^{-2}dx = \frac{-x}{x+4} + \int (x+4)^{-1}dx = \frac{-x}{x+4} + \ln|x+4| + C.$$

17. Using integration by parts $f(x) = \ln x^3$, $f'(x) = \dfrac{3}{x}dx$, $g'(x) = dx$, and $g(x) = x$. Therefore, $\displaystyle\int \ln x^3 dx = x\ln x^3 - \int 3dx = x\ln x^3 - 3x + C.$

19. Here, the first term requires an integration by parts, the second, a u-substitution, and the last, the basic rules for integration.
For first term, let $f(x) = 3x$, $f'(x) = 3dx$, $g'(x) = e^x dx$, and $g(x) = e^x$.
For u-substitution, let $u = x^2$ and $3du = 6xdx$. Therefore,

$$\int (3xe^x + 6xe^{x^2} + e^{5x})dx = \left[3xe^x - \int 3e^x dx\right]$$

$$+ \left[\int e^u (3du)\right] + \int e^{5x}dx$$

$$= 3xe^x - 3e^x + 3e^{x^2} + \frac{e^{5x}}{5} + C.$$

EXERCISES 7.3

1. Use $u = x^3$, so $du = (3x^2)dx$. The limits of integration are $(1)^3$ and $(4)^3$ or 1 and 64. Therefore,

$$\int_1^4 3x^2 e^{x^3}dx = \int_1^{64} e^u du = e^u\Big|_1^{64} = e^{64} - e$$

3. Using integration by parts, $f(x) = x$, $f'(x) = 1dx$, $g'(x) = (x-1)^{-1/2}$, and $g(x) = 2(x-1)^{1/2}$. Therefore,

$$\int_2^5 \left(\frac{x}{\sqrt{x-1}}\right) dx = 2x(x-1)^{1/2}\Big|_2^5 - \int_2^5 2(x-1)^{1/2} dx$$

$$= 2x(x-1)^{1/2}\Big|_2^5 - \frac{4}{3}(x-1)^{3/2}\Big|_2^5$$

$$= [10(2) - 4(1)] - \left[\frac{4}{3}(8) - \frac{4}{3}(1)\right] = \frac{20}{3}.$$

5. Use $u = x^4$, so $2du = 8x^3dx$, and the limits of integration are $(1)^4 = 1$ and $(2)^4 = 16$. Therefore,

$$\int_1^2 8x^3e^{x^4} dx = \int_1^{16} e^u(2du) = 2\ e^u\Big|_1^{16} = 2e^{16} - 2e$$

7. Use $u = x^2 + 144$, so $4du = 8xdx$, and the limits of integration are $(0)^2 + 144 = 144$ and $(5)^2 + 144 = 169$. Therefore,

$$\int_0^5 8x\sqrt{x^2+144}\, dx = \int_{144}^{169} u^{1/2}(4du) = \frac{8}{3}u^{3/2}\Big|_{144}^{169}$$

$$= \frac{8}{3}(13)^3 - \frac{8}{3}(12)^3 = \frac{3752}{3}.$$

9. Use $u = x^2$, so $du = 2xdx$, and $(1)^2 = 1$ and $(2)^2 = 4$ as the limits of integration for the first term. Use basic integration for the other terms. Therefore,

$$\int_1^2 (2xe^{x^2} + 4x + 3)\ dx = \int_1^4 e^u du + \int_1^2 (4x+3)dx$$

$$= e^u\Big|_1^4 + (2x^2 + 3x)\Big|_1^2$$

$$= (e^4 - e) + (8 + 6) - (2 + 3)$$

$$= e^4 - e + 9$$

11. Use $u = x^3 + 2x^2 + 1$, so $3du = (9x^2 + 12x)dx$, and the limits of integration are $(0)^3 + 2(0)^2 + 1 = 1$ and $(1)^3 + 2(1)^2 + 1 = 4$. Therefore,

$$\int_0^1 (9x^2 + 12x)(x^3 + 2x^2 + 1)^3 dx = \int_1^4 u^3(3du)$$
$$= \left.\frac{3u^4}{4}\right|_1^4 = \frac{768}{4} - \frac{3}{4} = \frac{765}{4}.$$

13. Using integration by parts, $f(x) = \ln x, f'(x) = \frac{1}{x}dx$, $g'(x) = x^4$, and $g(x) = \frac{x^5}{5}$. Therefore,

$$\int_1^e x^4 \ln x \; dx = \left.\frac{x^5}{5}\ln x\right|_1^e - \int_1^e \frac{x^4}{5}dx = \left.\frac{x^5}{5}\ln x\right|_1^e - \left.\frac{x^5}{25}\right|_1^e$$
$$= \left(\frac{e^5}{5} - 0\right) - \left(\frac{e^5}{25} - \frac{1}{25}\right) = \frac{4e^5}{25} + \frac{1}{25}.$$

15. Using integration by parts, $f(x) = 6x$, $f'(x) = 6dx$, $g'(x) = (x + 3)^{-3}$, and $g(x) = \frac{(x+3)^{-2}}{-2}$. Therefore,

$$\int_1^5 6x(x+3)^{-3}\; dx = -3x(x+3)^{-2}|_1^5 + \int_1^5 3(x+3)^{-2}dx$$
$$= -3x(x+3)^{-2}|_1^5 - \left.\frac{3}{(x+3)}\right|_1^5$$
$$= \left(\frac{-15}{64} + \frac{3}{16}\right) - \left(\frac{3}{8} - \frac{3}{4}\right) = \frac{21}{64}.$$

EXERCISES 7.4

1. Using partial fractions, $\frac{A}{x-1} + \frac{B}{x+1} = \frac{2}{x^2-1}$. Solving, $A(x+1) + B(x-1) = 2$ When $x = 1$, $2A = 2$ so $A = 1$ and when $x = -1$, $-2B = 2$ so $B = -1$. Therefore,

$$\int \left(\frac{2}{x^2-1}\right) dx = \int \left(\frac{1}{x-1} + \frac{1}{x+1}\right)dx = \ln|x-1| - \ln|x+1| + C.$$

3. Since the numerator is not of a lesser degree than the denominator, a long division reduces the integrand to $x + 3 + \frac{x+1}{x^2-1}$.
Using partial fractions, $\frac{A}{x-1} + \frac{B}{x+1} = \frac{x+1}{x^2-1}$.
Solving, $A(x+1) + B(x-1) = x + 1$
When $x = 1$, $2A = 2$ and $A = 1$ and when $x = -1$, $-2B = 0$ so $B = 0$ (note that the result is the same as reducing the fraction algebraically). Therefore,

$$\int \left(\frac{x^3 + 3x^2 - 2}{x^2 - 1}\right) dx = \int \left(x + 3 + \frac{1}{x-1}\right) dx$$

$$= \frac{x^2}{2} + 3x + \ln|x - 1| + C.$$

5. Using partial fractions, $\frac{A}{x-3} + \frac{B}{x+3} = \frac{1}{x^2-9}$.
Solving, $A(x+3) + B(x-3) = 1$ When $x = 3$, $6A = 1$ and $A = 1/6$ and when $x = -3$, $-6B = 1$ so $B = -1/6$. Therefore,

$$\int \frac{dx}{x^2 - 9} = \int \left[\frac{1}{6}\left(\frac{1}{x-3}\right) - \frac{1}{6}\left(\frac{1}{x+3}\right)\right] dx$$

$$= \frac{1}{6} \ln|x - 3| - \frac{1}{6} \ln|x + 3| + C.$$

7. Using partial fractions, $\frac{A}{x} + \frac{B}{x-1} + \frac{C}{x-2} = \frac{x^2+2}{x^3 - 3x^2 + 2x}$. Solving,
$A(x-1)(x-2) + B(x)(x-2) + C(x)(x-1) = x^2 + 2$.
When $x = 0$, $2A = 2$ and $A = 1$, when $x = 1$, $-B = 3$ so $B = -3$, and when $x = 2$, $2C = 6$ so $C = 3$. Therefore,

$$\int \left(\frac{x^2 + 2}{x^3 - 3x^2 + 2x}\right) dx = \int \left[\frac{1}{x} - \frac{3}{x-1} + \frac{3}{x-2}\right] dx$$

$$= \ln|x| - 3\ln|x - 1| + 3\ln|x - 2| + C$$

9. Using partial fractions, $\frac{A}{x} + \frac{B}{x+2} + \frac{C}{x-1} = \frac{6x^2+7x-4}{x^3+x^2-2x}$. Solving, $A(x+2)(x-1) + B(x)(x-1) + C(x)(x+2) = 6x^2 + 7x - 4$. When $x = 0$, $-2A = -4$ and $A = 2$, when $x = -2$, $6B = 6$ so $B = 1$, and when $x = 1$, $3C = 9$ so $C = 3$. Therefore,

$$\int \left(\frac{6x^2+7x-4}{x^3+x^2-2x}\right) dx = \int \left[\frac{2}{x} + \frac{1}{x+2} + \frac{3}{x-1}\right] dx$$

$$= 2\ln|x| + \ln|x+2| + 3\ln|x-1| + C.$$

EXERCISES 7.5

1. Here, $a = 2, b = 5, n = 6$, and $\Delta x = \frac{5-2}{6} = \frac{1}{2}$. The subintervals are, [2, 2.5], [2.5, 3], [3, 3.5], [3.5, 4], [4, 4.5], and [4.5, 5]. The six midpoints are 2.25, 2.75, 3.25, 3.75, 4.25, and 4.75.

3. Here, $a = 2, b = 5, n = 6$, and $\Delta x = \frac{5-2}{6} = \frac{1}{2}$. The endpoints are 2, 2.5, 3, 3.5, 4, 4.5, and 5.

5. Here, $a = 0$, $b = 4$, $n = 4$, and $\Delta x = \frac{4-0}{4} = 1$. The four midpoints are 0.5, 1.5, 2.5, and 3.5. Using the midpoint rule,

$$\int_0^4 (x^2+5)\, dx \approx 1[f(0.5) + f(1.5) + f(2.5) + f(3.5)]$$

$$= 5.25 + 7.25 + 11.25 + 17.25 = 41.$$

7. Here, $a = 1, b = 9, n = 8$, and $\Delta x = \frac{9-1}{8} = 1$. The eight midpoints are 1.5, 2.5, 3.5, 4.5, 5.5, 6.5, 7.5, and 8.5. Using the midpoint rule,

$$\int_1^9 (x^2+9x+8)\, dx \approx 1[f(1.5) + f(2.5) + f(3.5) + f(4.5)$$

$$+ f(5.5) + f(6.5) + f(7.5) + f(8.5)]$$

$$= [23.75 + 36.75 + 51.75 + 68.75$$

$$+ 87.75 + 108.75 + 131.75 + 156.75] = 666.$$

9. Here, $a = 0, b = 3, n = 6$, and $\Delta x = \dfrac{3-0}{6} = \dfrac{1}{2}$. Using the trapezoidal rule,

$$\int_0^3 (x^3 + 5x + 4)dx \approx \frac{1}{4}\left[f(0) + 2f\left(\frac{1}{2}\right) + 2f(1) + 2f\left(\frac{3}{2}\right)\right.$$
$$\left. +2f(2) + 2f\left(\frac{5}{2}\right) + f(3)\right]$$
$$= \frac{1}{4}[4 + 13.25 + 20 + 29.75 + 44 + 64.25 + 46]$$
$$= 55.31$$

11. Here, $a = 0, b = 4, n = 4$, and $\Delta x = \dfrac{4-0}{4} = 1$. Using Simpson's rule,

$$\int_0^4 (x^2 + 5)\,dx \approx \frac{1}{6}\left[f(0) + 4f(0.5) + 2f(1) + 4f(1.5) + 2f(2)\right.$$
$$\left. +4f(2.5) + 2f(3) + 4f(3.5) + f(4)\right]$$
$$= \frac{1}{6}\left[5 + 4(5.25) + 2(6) + 4(7.25) + 2(9)\right.$$
$$\left. +4(11.25) + 2(14) + 4(17.25) + 21 = 41.\overline{3}\right]$$

13. Here, $a = 1, b = 9, n = 8$, and $\Delta x = \dfrac{9-1}{8} = 1$. Using Simpson's rule,

$$\int_1^9 (x^2 + 9x + 8)dx \approx \frac{1}{6}\begin{bmatrix} f(1) + 4f(1.5) + 2f(2) + 4f(2.5) \\ +2f(3) + 4f(3.5) + 2f(4) + 4f(4.5) \\ +2f(5) + 4f(5.5) + 2f(6) + 4f(6.5) \\ +2f(7) + 4(f7.5) + 2f(8) + 4f(8.5) \\ +f(9) \end{bmatrix}$$
$$= \frac{1}{6}\begin{bmatrix} 18 + 95 + 60 + 147 + 88 + 207 \\ +120 + 275 + 156 + 351 + 196 \\ +435 + 240 + 527 + 288 + 627 + 170 \end{bmatrix}$$
$$= \frac{4000}{6} = 666.\overline{6}.$$

15. Here, $a = 2$, $b = 6$, $n = 4$, and $\Delta x = \dfrac{6-2}{4} = 1$. Using the trapezoidal rule,

$$\begin{aligned}\int_2^6 f(x)dx &\approx \frac{1}{2}[f(2) + 2f(3) + 2f(4) + 2f(5) + f(6)]\\ &= \frac{1}{2}[4.1 + 2(5.7) + 2(7.4) + 2(4.5) + 7.8]\\ &= \frac{1}{2}[47.1] = 23.55.\end{aligned}$$

EXERCISES 7.6

1. $$\begin{aligned}\int_5^\infty \frac{1}{x-3}dx &= \lim_{t\to\infty}\int_5^t \frac{1}{x-3}dx\\ &= \lim_{t\to\infty}[\ln|t-3| - \ln 2] = \infty. \text{ (diverges)}\end{aligned}$$

3. $$\int_1^\infty \frac{2}{x^4}dx = \lim_{t\to\infty}\int_1^t 2x^{-4}dx = \lim_{t\to\infty}\left[\frac{-2}{3t^3} + \frac{2}{3}\right] = \frac{2}{3}.$$

5. $$\int_{-\infty}^4 e^{4x}dx = \lim_{t\to-\infty}\int_t^4 e^{4x}dx = \lim_{t\to-\infty}\left[\frac{e^{16}}{4} - \frac{e^{4t}}{4}\right] = \frac{e^{16}}{4}.$$

7. $$\begin{aligned}\int_1^\infty e^{3x+1}dx &= \lim_{t\to\infty}\int_1^t e^{3x+1}dx\\ &= \lim_{t\to\infty}\left[\frac{e^{3t+1}}{3} - \frac{e^4}{3}\right] = \infty. \text{ (diverges)}\end{aligned}$$

9. $$\int_1^\infty 2x\ e^{-x^2}dx = \lim_{t\to\infty}\int_1^t 2xe^{-x^2}dx = \lim_{t\to\infty}[-e^{-t^2} + e^{-1}] = e^{-1}.$$

11. $$\begin{aligned}\int_2^\infty \frac{1}{x\ln x}dx &= \lim_{t\to\infty}\int_2^t \frac{1}{x\ln x}dx\\ &= \lim_{t\to\infty}[\ln|\ln t| - \ln|\ln 2|] = \infty. \text{ (diverges)}\end{aligned}$$

13. $$\begin{aligned}\int_2^\infty \frac{x^2}{\sqrt{x^3-4}}dx &= \lim_{t\to\infty}\int_2^\infty \frac{(x^2)}{\sqrt{x^3-4}}dx\\ &= \lim_{t\to\infty}\left[\frac{2}{3}\left(t^3-4\right)^{1/2} - \frac{2}{3}(4)^{1/2}\right] = \infty. \text{ (diverges)}\end{aligned}$$

15.

$$\int_{-\infty}^{2} \frac{2}{(4-x)^3}\,dx = \lim_{t\to-\infty}\int_{t}^{2} \frac{2}{(4-x)^3}\,dx$$

$$= \lim_{t\to-\infty}\left[\frac{1}{(4-2)^2} - \frac{1}{(4-t)^2}\right] = \frac{1}{4}.$$

SUPPLEMENTARY EXERCISES CHAPTER 7

1. Let $u = x^3 + 9$, $du = 3x^2dx$, and $\frac{8}{3}du = xdx$. Therefore,

$$\int 8x^2\sqrt{x^3+9}\;dx = \int u^{1/2}\left(\frac{8}{3}du\right)$$

$$= \frac{16}{9}u^{3/2} + C = \frac{16}{9}(x^3+9)^{3/2} + C.$$

3. Let $u = x^4 + x^2 + 5$, and $du = (4x^3 + 2x)dx$. Therefore,

$$\int \frac{4x^3+2x}{x^4+x^2+5}\,dx = \int \frac{1}{u}du = \ln|u| + C = \ln|x^4+x^2+5| + C.$$

5. Let $f(x) = 5x, f'(x) = 5dx$, $g'(x) = (5x+4)^{-1/3}dx$, and $g(x) = \frac{3}{10}(5x+4)^{2/3}$. Therefore,

$$\int \frac{5x}{\sqrt[3]{5x+4}}dx = \frac{3}{2}x(5x+4)^{2/3} - \int\left(\frac{3}{2}\right)(5x+4)^{2/3}dx$$

$$= \frac{3}{2}x(5x+4)^{2/3} - \frac{9}{50}(5x+4)^{5/3} + C$$

7. Let $f(x) = 5x+1, f'(x) = 5dx$, $g'(x) = e^{3x}dx$, and $g(x) = \frac{e^{3x}}{3}$.

$$\int(5x+1)e^{3x}dx = (5x+1)\frac{e^{3x}}{3} - \int\frac{5}{3}e^{3x}dx = (5x+1)\frac{e^{3x}}{3} - \frac{5e^{3x}}{9} + C.$$

9. Let $u = 5x^2$ and $2du = 20xdx$ so

$$\int_{1}^{2} 20xe^{5x^2}\,dx = \int_{5}^{20} 2e^{u}du = 2e^{u}\big|_{5}^{20} = 2e^{20} - 2e^{5}.$$

11. Using partial fractions, $\dfrac{A}{x+1} + \dfrac{B}{(x+1)^2} + \dfrac{C}{x-5} = \dfrac{2x^2 - 25x - 33}{(x+1)^2(x-5)}$.
Solving, $A(x+1)(x-5) + B(x-5) + C(x+1)^2 = 2x^2 - 25x - 33$.
When $x = -1$, $-6B = 6$, and $B = 1$. When $x = 5$, $36C = -108$ so $C = -3$. Since $(A + C)x^2 = 2x^2$, $A = 5$. Therefore,

$$\int \frac{2x^2 - 25x - 33}{(x+1)^2(x-5)}\,dx = \int \left[\frac{5}{x+1} + \frac{1}{(x+1)^2} - \frac{3}{x-5}\right] dx$$

$$= 5\ln|x+1| - \frac{1}{x+1} - 3\ln|x-5| + C.$$

13. Here, $a = 2$, $b = 6$, $\Delta x = \dfrac{6-2}{4} = 1$, and $f(x) = e^x$. Using the midpoint rule yields

$$1[f(2.5) + f(3.5) + f(4.5) + f(5.5)] = e^{2.5} + e^{3.5} + e^{4.5} + e^{5.5}.$$

Using the trapezoidal rule yields

$$\frac{1}{2}[f(2) + 2f(3) + 2f(4) + 2f(5) + f(6)] = \frac{1}{2}e^2 + e^3 + e^4 + e^5 + \frac{1}{2}e^6.$$

Using Simpson's Rule yields

$$\frac{1}{6}[f(2) + 4f(2.5) + 2f(3) + 4f(3.5) + 2f(4)$$
$$+ 4f(4.5) + 2f(5) + 4f(5.5) + f(6)]$$
$$= \frac{1}{6}[e^2 + 4e^{2.5} + 2e^3 + 4e^{3.5} + 2e^4 + 4e^{4.5} + 2e^5 + 4e^{5.5} + e^6]$$

15. $\displaystyle\int_{-\infty}^{3} e^{2x+1}\,dx = \lim_{t\to-\infty} \int_{t}^{3} e^{2x+1} dx = \lim_{t\to-\infty} \left[\frac{e^7}{2} - \frac{e^{2t+1}}{2}\right] = \frac{e^7}{2}.$

CHAPTER 8

FUNCTIONS OF SEVERAL VARIABLES

EXERCISES 8.1

1. $f(2, 5) = 2(2) + 3(5) = 19, f(3, -1) = 2(3) + 3(-1) = 3$, and $f(4, -3) = 2(4) + 3(-3) = -1$.

3. $f(1, 0) = 4(1) + 3(0)^2 = 4, f(2, -1) = 4(2) + 3(-1)^2 = 11$, and $f(2, 1) = 4(2) + 3(1)^2 = 11$.

5. $f(1, 2) = (1)^2 + 3(1) + (2)^3 + 2(2) + 5 = 21$,
$f(0, 1) = (0)^2 + 3(0) + (1)^3 + 2(1) + 5 = 8$, and
$f(-1, -2) = (-1)^2 + 3(-1) + (-2)^3 + 2(-2) + 5 = -9$.

7. $f(2, 0) = (2) + 2(0)^3 + e^0 = 3, f(0, 1) = (0) + 2(1)^3 + e^1 = 2 + e$, and $f(3, 0) = (3) + 2(0)^3 + e^0 = 4$.

9. $f(0, 1, 2) = (0)^2 + 3(1) + (2)^3 = 11$,
$f(0.5, 1, 1) = (0.5)^2 + 3(1) + (1)^3 = 4.25$, and
$f(-1, 0, 2) = (-1)^2 + 3(0) + (2)^3 = 9$.

11.
$$f(3 + h, 4) - f(3, 4) = [(3 + h)^2 + 4^2] - (3^2 + 4^2)$$
$$= (9 + 6h + h^2 + 16] - (25) = h^2 + 6h$$

Solutions Manual to Accompany Fundamentals of Calculus, First Edition. Carla C. Morris and Robert M. Stark.

13. $f(2a, 2b) = 7(2a)^{2/3}(2b)^{1/3} = 7(2)^{2/3}(2)^{1/3}a^{2/3}b^{1/3}$
$= 2(7a^{2/3}b^{1/3}) = 2f(a, b).$

15. The level curves for $f(x, y) = x + y$ and $c = 1, 4,$ and 9 are shown below:

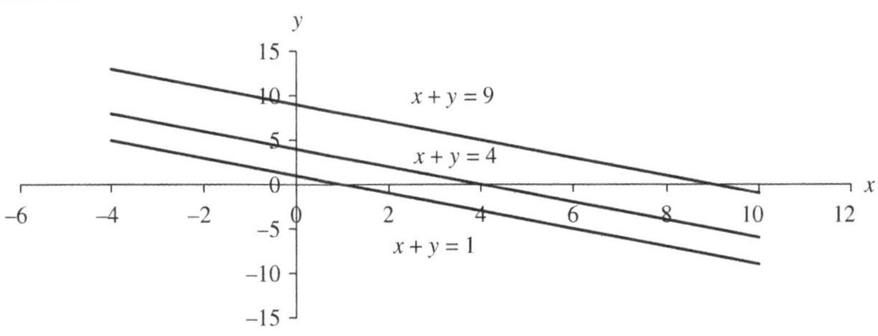

17. The level curves for $f(x, y) = xy$ and $c = 1, 4,$ and 9 are shown below:

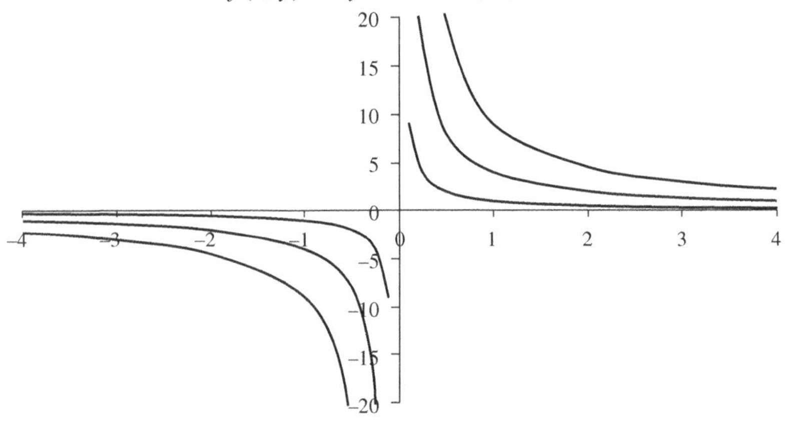

EXERCISES 8.2

1. $f_x = 20x^3 - 6$ and $f_y = -6y + 2$

3. $f_x = 25(5x^7 + 4y^5 + 7y + 3)^{24}(35x^6)$ and
$f_y = 25(5x^7 + 4y^5 + 7y + 3)^{24}(20y^4 + 7)$

5. $f_x = 3x^2y^5(e^{x^3y^5}) + 36x^3$ and $f_y = 5x^3y^4(e^{x^3y^5}) + 3$

7. $f_x = \dfrac{-y^5}{x^2}$ and $f_y = \dfrac{5y^4}{x}$

9. First, rewrite the function as $f(x, y) = (x^2y)^{1/3}$. The partial derivatives are $f_x = \frac{1}{3}(x^2y)^{-2/3}(2xy) = \frac{2}{3}x^{-1/3}y^{1/3}$ and
$f_y = \frac{1}{3}(x^2y)^{-2/3}(x^2) = \frac{1}{3}x^{2/3}y^{-2/3}$

11. Think of the function as the product $f(x, y) = (x^3e^x)(y^8)$,
$f_x = (x^3e^x + 3x^2e^x)y^8$, and $f_y = x^3e^x(8y^7)$

13. First, rewrite the function as $x^2(y - x)^{1/3}$,
$f_x = x^2\left[\frac{1}{3}(y - x)^{-2/3}(-1)\right] + (y - x)^{1/3}[2x]$ and
$f_y = \left[\frac{1}{3}(y - x)^{-2/3}(1)\right](x^2)$

15. $f_x = 15x^2, f_y = 4y$, and $f_z = 12z^3$

17. $f_x = 3x^2y^2 + 4ze^{4xz}, f_y = 5 + 2x^3y + 3z^4$, and $f_z = 12yz^3 + 4xe^{4xz}$

19. $f_x = 6xy^3 + 2$ and $f(4, 1) = 6(4)(1)^3 + 2 = 26$
$f_y = 9x^2y^2 + 18y^5$ and $f(4, 1) = 9(4)^2(1)^2 + 18(1)^5 = 162$

21. The marginal productivities are $f_x = 3x^{-1/4}y^{1/4}$ and $f_y = x^{3/4}y^{-3/4}$.

EXERCISES 8.3

1. The first partial derivatives are $f_x = 15x^2 - 4x$ and $f_y = 4$. The second partial derivatives are $f_{xx} = 30x - 4, f_{yy} = 0$, and $f_{xy} = f_{yx} = 0$.

3. The first partial derivatives are $f_x = 10xy + 9y^5$ and $f_y = 5x^2 + 45xy^4$. The second partial derivatives are $f_{xx} = 10y, f_{yy} = 180xy^3$, and $f_{xy} = f_{yx} = 10x + 45y^4$.

5. The first partial derivatives are $f_x = 3x^2y + 9y^4$ and $f_y = x^3 + 36xy^3$. The second partial derivatives are $f_{xx} = 6xy, f_{yy} = 108xy^2$, and $f_{xy} = f_{yx} = 3x^2 + 36y^3$.

7. The first partial derivatives are $f_x = \frac{-y^2}{x^2}$ and $f_y = \frac{2y}{x}$. The second partial derivatives are $f_{xx} = \frac{2y^2}{x^3}$, $f_{yy} = \frac{2}{x}$, and $f_{xy} = f_{yx} = \frac{-2y}{x^2}$.

9. The first partial derivatives are $f_x = 3x^2y^4e^{x^3y^4} + 32x^7 + 28x^3y^5$ and $f_y = 4x^3y^3e^{x^3y^4} + 35x^4y^4$. The second partial derivatives are
$f_{xx} = 3x^2y^4[3x^2y^4(e^{x^3y^4})] + (e^{x^3y^4})[6xy^4] + 224x^6 + 84x^2y^5$,
$f_{yy} = 4x^3y^3[4x^3y^3(e^{x^3y^4})] + (e^{x^3y^4})[12x^3y^2] + 140x^4y^3$, and
$f_{xy} = f_{yx} = 3x^2y^4[4x^3y^3(e^{x^3y^4})] + (e^{x^3y^4})[12x^2y^3] + 140x^3y^4$.

11. Possible extrema occur when the first partial derivatives $f_x = 2x$ and $f_y = -2y$ are set to zero. The only possible extremum is at (0, 0).

13. Possible extrema occur when the first partial derivatives $f_x = 2x + 8y$ and $f_y = 8x + 2y$ are set to zero. The only possible extremum is at (0, 0).

15. To determine possible local extrema for a bivariate function, the first partial derivatives are set to zero. Then, second partial derivatives determine D and the nature of the possible extrema.
$f_x = 2x + 2$, $f_y = 2y - 6$. Simultaneously setting these to zero yields $(-1, 3)$ as a possible extremum. Using $f_{xx} = 2$, $f_{yy} = 2$, and $f_{xy} = f_{yx} = 0$, $D(-1, 3) = (2)(2) - (0)^2 > 0$. Since $D(-1, 3)$ is positive and $f_{xx} > 0$, there is a local minimum $f(-1, 3) = -10$.

17. To determine possible local extrema for a bivariate function, the first partial derivatives are set to zero. Then, second partial derivatives determine D and the nature of the possible extrema. $f_x = 6x - 4y + 8$, $f_y = -4x + 6y - 17$. Simultaneously setting these to zero yields (1, 7/2) as a possible extremum.

Using $f_{xx} = 6$, $f_{yy} = 6$, and $f_{xy} = f_{yx} = -4$, $D(1, 7/2) = (6)(6) - (-4)^2 > 0$. Since $D(1, 7/2)$ is positive and $f_{xx} > 0$ there is a local minimum $f(1, 7/2) = 17/4$.

19. To determine possible local extrema for a bivariate, the first partial derivatives are set to zero. Then, second partial derivatives determine D and the nature of the possible extrema.

$f_x = 40 - 2x - y$, $f_y = 50 - 2y - x$. Simultaneously setting these to zero yields (10, 20) as a possible extremum.

Using $f_{xx} = -2$, $f_{yy} = -2$, and $f_{xy} = f_{yx} = -1$, $D(10, 20) = (-2)(-2) - (-1)^2 > 0$. Since $D(10, 20)$ is positive and $f_{xx} < 0$, there is a local maximum $f(10, 20) = 700$.

21. To determine possible local extrema for a bivariate function, the first partial derivatives are set to zero. Then, second partial derivatives determine D and the nature of the possible extrema.

$f_x = 6x^2 - 6$, $f_y = -4y + 8$. Simultaneously setting these to zero yields (1, 2) and $(-1, 2)$ as possible extrema.

Using $f_{xx} = 12x$, $f_{yy} = -4$, and $f_{xy} = f_{yx} = 0$, $D(1, 2) = (12)(-4) - (0)^2 < 0$ indicating no extremum at (1, 2). Next, $D(-1, 2) = (-12)(-4) - (0)^2 > 0$ is positive and $f_{xx} < 0$, there is a local maximum $f(-1, 2) = 27$.

23. To determine possible local extrema for a bivariate function, the first partial derivatives are set to zero. Then, the second partial derivatives determine D and the nature of the possible extrema.

$f_x = 3x^2 - 2y$, $f_y = -2x + 12y^3$. Simultaneously setting these to zero yields $x = 0$ or $x = \sqrt[5]{\frac{4}{81}}$. The possible extrema are (0, 0) and (0.5479, 0.4503).

Using $f_{xx} = 6x$, $f_{yy} = 36y^2$, and $f_{xy} = f_{yx} = -2$, $D(0, 0) = (6)(0) - (-2)^2 < 0$ indicating no extremum at (0, 0). Next, $D(0.5479, 0.4503) = (3.2874)(7.3) - (-2)^2 > 0$ and $f_{xx} > 0$ indicating a local minimum at $f(0.5479, 0.4503) = -0.2056$.

25. First, $xyz = 125$ and, $z = 125/xy$. One seeks to minimize $2xy + 2y\left(\frac{125}{xy}\right) + 2x\left(\frac{125}{xy}\right) = 2xy + \frac{250}{x} + \frac{250}{y}$. The first partial derivatives are $f_x = 2y - \frac{250}{x^2}$ and $f_y = 2x - \frac{250}{y^2}$. Setting these to zero yields the trivial solution, $x = 0$, which is not possible. The other possibility is that $x = 5$. The possible extrema occur when x and y (and also z) = 5. The second partial derivatives are $f_{xx} = \frac{500}{x^3}$, $f_{yy} = \frac{500}{y^3}$, and $f_{xy} = f_{yx} = 2$. $D = (4)(4) - (2)^2 > 0$ and $f_{xx} > 0$, so there is a minimum when a cube with edge 5 ft with 150 ft^2 of insulation.

27. The first partial derivatives are $f_x = 3 - y - 2x$ and $f_y = 4 - x - 4y$. Setting these to zero yields $\left(\frac{8}{7}, \frac{5}{7}\right)$ as a possible extremum. The second partial derivatives are $f_{xx} = -2$, $f_{yy} = -4$, and $f_{xy} = f_{yx} = -1$. $D = (-2)(-4) - (-1)^2 > 0$ and $f_{xx} < 0$. The maximum revenue is $f\left(\frac{8}{7}, \frac{5}{7}\right) = \frac{154}{49} = \frac{22}{7}$.

EXERCISES 8.4

1. First, $E = (6 - B - 3A)^2 + (9 - B - 4A)^2 + (15 - B - 5A)^2$. The first partial derivatives are

$$\begin{aligned} E_A &= 2(6 - B - 3A)(-3) + 2(9 - B - 4A)(-4) + 2(15 - B - 5A)(-5) \\ &= 100A + 24B - 258 \end{aligned}$$

$$E_B = 2(6 - B - 3A)(-1) + 2(9 - B - 4A)(-1) + 2(15 - B - 5A)(-1)$$
$$= 24A + 6B - 60$$

Solving the system

$$100A + 24B = 258$$
$$24A + 6B = 60$$

yields $A = 9/2$ and $B = -8$, so the least squares line is $\hat{y} = \frac{9}{2}x - 8$.

3. First, $E = (6 - B - 4A)^2 + (8 - B - 5A)^2 + (4 - B - 6A)^2$. The first partial derivatives are

$$E_A = 2(6 - B - 4A)(-4) + 2(8 - B - 5A)(-5) + 2(4 - B - 6A)(-6)$$
$$= 154A + 30B - 176$$
$$E_B = 2(6 - B - 4A)(-1) + 2(8 - B - 5A)(-1) + 2(4 - B - 6A)(-1)$$
$$= 30A + 6B - 36$$

Solving the system

$$154A + 30B = 176$$
$$30A + 6B = 36$$

yields $A = -1$ and $B = 11$, so the least squares line is $\hat{y} = -x + 11$.

5. First, $E = (5 - B)^2 + (11 - B - 4A)^2 + (18 - B - 8A)^2$. The first partial derivatives are

$$E_A = 2(11 - B - 4A)(-4) + 2(18 - B - 8A)(-8)$$
$$= 160A + 24B - 376$$
$$E_B = 2(5 - B)(-1) + 2(11 - B - 4A)(-1) + 2(15 - B - 8A)(-1)$$
$$= 24A + 6B - 68$$

Solving the system

$$160A + 24B = 376$$
$$24A + 6B = 68$$

yields $A = 13/8$ and $B = 29/6$, so the least squares line is $\hat{y} = \frac{13}{8}x + \frac{29}{6}$.

7. First,
$E = (12 - B - A)^2 + (11 - B - 2A)^2 + (9 - B - 3A)^2 + (6 - B - 4A)^2$.
The first partial derivatives are

$$\begin{aligned} E_A &= 2(12 - B - A)(-1) + 2(11 - B - 2A)(-2) \\ &\quad + 2(9 - B - 3A)(-3) + 2(6 - B - 4A)(-4) \\ &= 60A + 20B - 170 \\ E_B &= 2(12 - B - A)(-1) + 2(11 - B - 2A)(-1) \\ &\quad + 2(9 - B - 3A)(-1) + 2(6 - B - 4A)(-1) \\ &= 20A + 8B - 76 \end{aligned}$$

Solving the system

$$\begin{aligned} 60A + 20B &= 170 \\ 20A + 8B &= 76 \end{aligned}$$

yields $A = -2$ and $B = 29/2$, so the least squares line is
$\hat{y} = -2x + \dfrac{29}{2}$.

9. First,
$E = (10 - B - 3A)^2 + (27 - B - 4A)^2 + (52 - B - 5A)^2 + (102 - B - 6A)^2$. The first partial derivatives are

$$\begin{aligned} E_A &= 2(10 - B - 3A)(-3) + 2(27 - B - 4A)(-4) \\ &\quad + 2(52 - B - 5A)(-5) + 2(102 - B - 6A)(-6) \\ &= 172A + 36B - 2020 \\ E_B &= 2(10 - B - 3A)(-1) + 2(27 - B - 4A)(-1) \\ &\quad + 2(52 - B - 5A)(-1) + 2(102 - B - 6A)(-1) \\ &= 36A + 8B - 382 \end{aligned}$$

Solving the system

$$\begin{aligned} 172A + 36B &= 2020 \\ 36A + 8B &= 382 \end{aligned}$$

yields $A = 301/10$ and $B = -877/10$, so the least squares line is
$\hat{y} = \dfrac{301}{10}x - \dfrac{877}{10}$.

11. First,
$E = (9 - B - 2A)^2 + (7 - B - 3A)^2 + (6 - B - 4A)^2 + (10 - B - 5A)^2$.
The first partial derivatives are

$$\begin{aligned} E_A &= 2(9 - B - 2A)(-2) + 2(7 - B - 3A)(-3) \\ &\quad + 2(6 - B - 4A)(-4) + 2(10 - B - 5A)(-5) \\ &= 108A + 28B - 226 \\ E_B &= 2(9 - B - 2A)(-1) + 2(7 - B - 3A)(-1) \\ &\quad + 2(6 - B - 4A)(-1) + 2(10 - B - 5A)(-1) \\ &= 28A + 8B - 64 \end{aligned}$$

Solving the system

$$\begin{aligned} 108A + 28B &= 226 \\ 28A + 8B &= 64 \end{aligned}$$

yields $A = 1/5$ and $B = 73/10$, so the least squares line is
$\hat{y} = \frac{1}{5}x + \frac{73}{10}$.

13. First, $E = (5 - B - A)^2 + (9 - B - 3A)^2 + (11 - B - 4A)^2$.
The first partial derivatives are

$$\begin{aligned} E_A &= 2(5 - B - A)(-1) + 2(9 - B - 3A)(-3) + 2(11 - B - 4A)(-4) \\ &= 52A + 16B - 152 \\ E_B &= 2(5 - B - A)(-1) + 2(9 - B - 3A)(-1) + 2(11 - B - 4A)(-1) \\ &= 16A + 6B - 50 \end{aligned}$$

Solving the system

$$\begin{aligned} 52A + 16B &= 152 \\ 16A + 6B &= 50 \end{aligned}$$

yields $A = 2$ and $B = 3$, so the least squares line is $\hat{y} = 2x + 3$.
The least squares error is
$E = (5 - 2 - 3)^2 + (9 - 3 - 3(2))^2 + (11 - 4(2) - 3)^2 = 0$. Since the points are collinear, there is no estimation error.

The points (1, 5), (3, 9), and (4, 11) lie on the line $y = 2x + 3$ and, therefore, it is the least squares line.

15. First, $E = (25 - B - A)^2 + (30 - B - 4A)^2 + (32 - B - 5A)^2$.
The first partial derivatives are

$$E_A = 2(25 - B - A)(-1) + 2(30 - B - 4A)(-4) + 2(32 - B - 5A)(-5) = 84A + 20B - 610$$

$$E_B = 2(25 - B - A)(-1) + 2(30 - B - 4A)(-1) + 2(32 - B - 5A)(-1) = 20A + 6B - 174$$

Solving the system

$$84A + 20B = 610$$
$$20A + 6B = 174$$

yields $A = 45/26$ and $B = 302/13$, so the least squares line is $\hat{y} = \frac{45}{26}x + \frac{302}{13}$.

The price 2 months after time t is \$26.69.

EXERCISES 8.5

1. To maximize $x^2 - y^2$ subject to the constraint $2x + y = 3$, one writes the Lagrangian as

$$F(x, y, \lambda) = x^2 - y^2 + \lambda(3 - 2x - y)$$

The partial derivatives with respect to x and y are

$$F_x = 2x - 2\lambda$$
$$F_y = -2y - \lambda$$

Setting these to zero yields that $x = -2y$. Using the partial derivative with respect to λ yields $F_\lambda = 3 - 2x - y$. Setting this derivative to zero and using the relationship $x = -2y$ yields $x = 2$ and $y = -1$, so the maximum of $x^2 - y^2$ is $(2)^2 - (-1)^2 = 3$.

3. To maximize xy subject to the constraint $x + y = 16$, the Lagrangian is

$$F(x, y, \lambda) = xy + \lambda(16 - x - y)$$

The partial derivatives with respect to x and y are

$$F_x = y - \lambda$$
$$F_y = x - \lambda$$

Setting these first partials to zero yields $x = y$. The partial derivative with respect to λ yields $F_\lambda = 16 - x - y$. Setting this derivative to zero and using the relationship $x = y$ yields $x = 8$ and $y = 8$, so the maximum of xy is $(8)(8) = 64$.

5. For the point on the parabola $y^2 = 4x$ that is closest to (1, 0) minimize the squared distance $(x - 1)^2 + (y - 0)^2$ subject to $y^2 = 4x$. The Lagrangian is

$$F(x, y, \lambda) = (x - 1)^2 + y^2 + \lambda(4x - y^2)$$

The partial derivatives with respect to x and y are

$$F_x = 2(x - 1) + 4\lambda$$
$$F_y = 2y - 2y\lambda$$

Focusing on F_y yields that $y = 0$ or $\lambda = 1$. If the latter is true, then $x = -1$, which is not possible

$$(F_\lambda = 4x - y^2).$$

Therefore, $y = 0$ and then $x = 0$. The point (0, 0) is the point on the parabola $y^2 = 4x$ that is closest to (1, 0).

7. To find two positive numbers that add to 9 that maximize x^2y, the Lagrangian

$$F(x, y, \lambda) = x^2y + \lambda(9 - x - y)$$

The partial derivatives with respect to x and y are

$$F_x = 2xy - \lambda$$
$$F_y = x^2 - \lambda$$

⟶

The partial derivatives indicate that $x = 0$ or that $x = 2y$. If $x = 0$, the product x^2y is a minimum not a maximum. Therefore, using $x = 2y$ and $F_\lambda = 9 - x - y$, one determines that $x = 6$ and $y = 3$ and that the maximum value of x^2y is 108.

9. Here, one seeks to maximize xy subject to the constraint $16x + 10y = 280$ (y is the length of the east and west sides). Forming the Lagrangian, one has

$$F(x, y, \lambda) = xy + \lambda(280 - 16x - 10y)$$

The partial derivatives with respect to x and y are

$$F_x = y - 16\lambda$$
$$F_y = x - 10\lambda$$

These partial derivatives indicate that $y = (8/5)x$. Substituting this relationship into the partial derivative $F_\lambda = (280 - 16x - 10y)$ yields $x = 8.75$ ft and $y = 14$ ft.

11. To minimize $2x^2 + 4xy$ subject to the constraint $x^2y = 64{,}000$. The Lagrangian is $F(x, y, \lambda) = 2x^2 + 4xy + \lambda(64{,}000 - x^2y)$

The partial derivatives with respect to x and y are

$$F_x = 4x + 4y - 2xy\lambda$$
$$F_y = 4x - x^2\lambda$$

The partial derivative with respect to y has $x = 0$ and discarded or $x = 4$. Substituting this information into the partial derivative with respect to x yields that $y = \dfrac{4}{\lambda}$. Therefore, $x = y$. Using $F_\lambda = (64,000 - x^2y), x = 40$ in. and $y = 40$ in. forming a cube with the least amount of material.

EXERCISES 8.6

1. First, evaluate $\int_{-1}^{2} (5x - 2y)dy$ to yield

$(5xy - y^2)|_{-1}^{2} = (10x - 4) - (-5x - 1) = 15x - 3.$

Therefore,

$$\int_0^3 \left[\int_{-1}^2 (5x-2y)\,dy\right] dx = \int_0^3 (15x-3)dx$$

$$= \left(\frac{15x^2}{2} - 3x\right)\Big|_0^3 = \frac{117}{2}.$$

3. First, evaluate $\int_{-2}^{3} (6xy^2 - x^4)dy$ to yield

$(2xy^3 - x^4y)|_{-2}^{3} = (54x - 3x^4) - (-16x + 2x^4) = 70x - 5x^4.$

Therefore,

$$\int_1^3 \left[\int_{-2}^3 (6xy^2 - x^4)dy\right] dx = \int_1^3 (70x - 5x^4)dx = (35x^2 - x^5)|_1^3 = 38.$$

5. First, evaluate $\int_0^2 (x^4y^5)dy$ to yield $\left(\frac{x^4y^6}{6}\right)\Big|_0^2 = \frac{32x^4}{3}.$

Therefore,

$$\int_1^4 \left[\int_0^2 x^4y^5dy\right] dx = \int_1^4 \left(\frac{32x^4}{3}\right)dx = \left(\frac{32x^5}{15}\right)\Big|_1^4 = \frac{10{,}912}{5}.$$

7. First, evaluate $\int_0^x (x+y+2)dy$ to yield

$\left(xy + \frac{y^2}{2} + 2y\right)\Big|_0^x = \left(\frac{3}{2}x^2 + 2x\right).$

Therefore,

$$\int_0^2 \left[\int_0^x (x+y+2)dy\right] dx = \int_0^2 \left(\frac{3x^2}{2} + 2x\right)dx = \left(\frac{x^3}{2} + x^2\right)\Big|_0^2 = 8.$$

9. First, evaluate $\int_0^{x/2} (e^{2y-x})dy$ to yield $\left(\frac{e^{2y-x}}{2}\right)\Big|_0^{x/2} = \left(\frac{1}{2} - \frac{e^{-x}}{2}\right).$

Therefore,

$$\int_0^3 \left[\int_0^{x/2} \left(e^{2y-x}\right) dy\right] dx = \int_0^3 \left(\frac{1}{2} - \frac{e^{-x}}{2}\right)dx$$

$$= \left(\frac{1}{2}x + \frac{e^{-x}}{2}\right)\Big|_0^3 = 1 + \frac{e^{-3}}{2}.$$

11. First, evaluate $\int_{y^2}^{y-1} (5x-2y)dx$ to yield

$\left(\frac{5x^2}{2} - 2xy\right)\Big|_{y^2}^{y-1} = -\frac{5y^4}{2} + 2y^3 + \frac{y^2}{2} - 3y + \frac{5}{2}.$

⟶

Therefore,

$$\int_0^1 \left[\int_{y^2}^{y-1} (5x - 2y)\,dx\right] dy = \int_0^1 \left(-\frac{5y^4}{2} + 2y^3 + \frac{y^2}{2} - 3y + \frac{5}{2}\right) dy$$

$$= \left(\frac{-y^5}{2} + \frac{y^4}{2} + \frac{y^3}{6} - \frac{3y^2}{2} + \frac{5}{2}y\right)\Bigg|_0^1 = \frac{7}{6}.$$

SUPPLEMENTARY EXERCISES CHAPTER 8

1. The first partial derivatives are $f_x = 7x^6y^5e^{x^7y^5} + 36x^3y^3 + 32$ and $f_y = 5x^7y^4e^{x^7y^5} + 27x^4y^2$

3. $f(mx, my) = k(mx)^\alpha(my)^\beta = km^{\alpha+\beta}x^\alpha y^\beta = m^{\alpha+\beta}\lfloor kx^\alpha y^\beta \rfloor$

$$= m^1 \lfloor kx^\alpha y^\beta \rfloor = m(f(x, y)).$$

5. $f_x = 10(2x^3 - 3x + y^4)^9[6x^2 - 3]$ and $f_y = 10(2x^3 - 3x + y^4)^9[4y^3]$

7. The first partial derivatives are $f_x = 3x^2 - 2xy + y^2$ and $f_y = -x^2 + 2xy - 3y^2$. The second partial derivatives are

$f_{xx} = 6x - 2y,\ \ f_{yy} = 2x - 6y,\ \ f_{xy} = -2x + 2y,$ and $f_{yx} = -2x + 2y.$
The mixed derivatives f_{xy} and f_{yx} are equal.

9. First, $E = (7 - B - 3A)^2 + (15 - B - 4A)^2 + (23 - B - 7A)^2$.
The first partial derivatives are

$$E_A = 2(7 - B - 3A)(-3) + 2(15 - B - 4A)(-4)$$
$$+ 2(23 - B - 7A)(-7)$$
$$= 148A + 28B - 484$$

$$E_B = 2(7 - B - 3A)(-1) + 2(15 - B - 4A)(-1)$$
$$+ 2(23 - B - 7A)(-1)$$
$$= 28A + 6B - 90$$

Solving the system

$$148A + 28B = 484$$
$$28A + 6B = 90$$

yields $A = 48/13$ and $B = -29/13$, so the least squares line is $\hat{y} = \frac{48}{13}x - \frac{29}{13}$.

11. Using $E = \sum_{i=1}^{n}(y_i - \beta_0 - \beta_1 x_i)^2$, the first partial derivatives are

$$E_{\beta_1} = 2(y_1 - \beta_0 - \beta_1 x_1)(-x_1) + 2(y_2 - \beta_0 - \beta_1 x_2)(-x_2) + \cdots + 2(y_n - \beta_0 - \beta_1 x_n)(-x_n)$$

$$E_{\beta_0} = 2(y_1 - \beta_0 - \beta_1 x_1)(-1) + 2(y_2 - \beta_0 - \beta_1 x_2)(-1) + \cdots + 2(y_n - \beta_0 - \beta_1 x_n)(-1)$$

rearranging terms yields

$$2\beta_1(x_1^2 + x_2^2 + \cdots x_n^2) + 2\beta_0(x_1 + x_2 + \cdots + x_n) = 2(x_1 y_1 + x_2 y_2 + \cdots + x_n y_n)$$

and

$$2\beta_1(x_1 + x_2 + \cdots + x_n) + 2n\beta_0 = 2(y_1 + y_2 + \cdots + y_n)$$

Solving the system of equations as

$$n\left[2\beta_1 \sum x^2 + 2\beta_0 \sum x = 2\sum xy\right]$$

$$\left(-\sum x\right)\left[2\beta_1 \sum x + 2n\beta_0 = 2\sum y\right]$$

yields $n\beta_1 \sum x^2 - \beta_1\left(\sum x\right)^2 = n\sum xy - \sum x \sum y$ and, therefore,

$$\beta_1 = \frac{n\sum xy - \sum x \sum y}{n\sum x^2 - \left(\sum x\right)^2} = \frac{\sum xy - \dfrac{\sum x \sum y}{n}}{\sum x^2 - \dfrac{\left(\sum x\right)^2}{n}} = \frac{SS_{xy}}{SS_{xx}}$$

and $\beta_0 = \dfrac{\sum y}{n} - \beta_1 \dfrac{\sum x}{n} = \bar{y} - \beta_1 \bar{x}$

13. Here, to minimize $F(x, y) = x^2 + y^2$ subject to $x + y = f$, form the Lagrangian as $F(x, y, \lambda) = x^2 + y^2 + \lambda(f - x - y)$. The partial derivatives are $F_x = 2x - \lambda$ and $F_y = 2y - \lambda$ indicating that $\lambda = 2x = 2y$ or that $x = y$. To minimize the sum of squares of two numbers which add to a fixed number, the numbers equal $f/2$.

15. First, evaluate $\int_0^1 e^{x+y}dx = (e^{x+y})|_0^1 = e^{l+y} - e^y$. Next,

$$\begin{aligned}\int_1^2 \left[\int_0^1 e^{x+y}dx\right] dy &= \int_1^2 e^{1+y} - e^y)dy \\ &= (e^{1+y} - e^y)|_1^2 = (e^3 - e^2) - (e^2 - e) \\ &= e^3 - 2e^2 + e.\end{aligned}$$

CHAPTER 9

SERIES AND SUMMATIONS

EXERCISES 9.1

1. Here, $a = 1$ and $r = \frac{1}{2}$. Therefore, $S = \dfrac{1}{1 - \frac{1}{2}} = 2$.

3. Here, $a = \frac{2}{3}$ and $r = \frac{1}{3}$. Therefore, $S = \dfrac{\frac{2}{3}}{1 - \frac{1}{3}} = 1$.

5. Here, $a = \frac{4}{5}$ and $r = \frac{-1}{5}$. Therefore, $S = \dfrac{\frac{4}{5}}{1 - \left(\frac{-1}{5}\right)} = \frac{2}{3}$.

7. Here, $a = \frac{5}{2}$ and $r = \frac{1}{2}$. Therefore, $S = \dfrac{\frac{5}{2}}{1 - \frac{1}{2}} = 5$.

9. Here, $a = 5$ and $r = \frac{-1}{4}$. Therefore, $S = \dfrac{5}{1 - \left(\frac{-1}{4}\right)} = 4$.

Solutions Manual to Accompany Fundamentals of Calculus, First Edition. Carla C. Morris and Robert M. Stark.

11. Here, 0.11111 … is rewritten as $\frac{1}{10} + \frac{1}{10^2} + \frac{1}{10^3} + \cdots$

Therefore, $a = \frac{1}{10}$ and $r = \frac{1}{10}$ and $S = \frac{\frac{1}{10}}{1 - \frac{1}{10}} = \frac{1}{9}$.

13. Here, 0.161616 … is rewritten as $\frac{16}{10^2} + \frac{16}{10^4} + \frac{16}{10^6} + \cdots$

Therefore, $a = \frac{16}{100}$ and $r = \frac{1}{100}$ and $S = \frac{\frac{16}{100}}{1 - \frac{1}{100}} = \frac{16}{99}$.

15. Here, 0.135135 … is rewritten as $\frac{135}{10^3} + \frac{135}{10^6} + \frac{135}{10^9} + \cdots$

Therefore, $a = \frac{135}{1000}$ and $r = \frac{1}{1000}$ and

$$S = \frac{\frac{135}{1000}}{1 - \frac{1}{1000}} = \frac{135}{999} = \frac{5}{37}.$$

17. Here, 1.44444 … is rewritten as $(1) + \frac{4}{10} + \frac{4}{10^2} + \frac{4}{10^3} + \cdots$

Therefore, $a = \frac{4}{10}$ and $r = \frac{1}{10}$ and

$$S = (1) + \frac{\frac{4}{10}}{1 - \frac{1}{10}} = (1) + \frac{4}{9} = \frac{13}{9}.$$

19. Here, 3.4343 … is rewritten as $(3) + \frac{43}{10^2} + \frac{43}{10^4} + \frac{43}{10^6} + \cdots$

Therefore, $a = \frac{43}{100}$ and $r = \frac{1}{100}$ and

$$S = (3) + \frac{\frac{43}{100}}{1 - \frac{1}{100}} = (3) + \frac{43}{99} = \frac{340}{99}.$$

21. $\sum_{x=3}^{\infty} \left(\frac{1}{3}\right)^x = \frac{1}{3^3} + \frac{1}{3^4} + \frac{1}{3^5} + \cdots$ Therefore, $a = \frac{1}{27}$ and $r = \frac{1}{3}$ and

$$S = \frac{\frac{1}{27}}{1 - \frac{1}{3}} = \frac{1}{18}.$$

23. $\sum_{x=1}^{\infty}\left(\frac{3}{5}\right)^x = \frac{3}{5} + \frac{3^2}{5^2} + \frac{3^3}{5^3} + \cdots$ Therefore, $a = \frac{3}{5}$ and $r = \frac{3}{5}$ and

$$S = \frac{\frac{3}{5}}{1 - \frac{3}{5}} = \frac{3}{2}.$$

25. For the sum
$15 + 15(0.80) + 15(0.80)^2 + 15(0.80)^3 + \cdots$
$a = 15$ and $r = 0.2$, so $S = \dfrac{15}{1 - 0.80} = 75.$
There is a 75 million dollar multiplier effect by introducing 15 million dollars to the economy.

EXERCISES 9.2

1. The function and the first three derivatives are
$f(x) = e^x, f'(x) = e^x, f''(x) = e^x$, and $f'''(x) = e^x$.
Evaluated at $x = 0$:
$f(0) = 1, f'(0) = 1,\ f''(0) = 1$, and $f'''(0) = 1$.
The third-degree Maclaurin Polynomial is

$$1 + 1(x) + 1\left(\frac{x^2}{2!}\right) + 1\left(\frac{x^3}{3!}\right) = 1 + x + \frac{x^2}{2} + \frac{x^3}{6}.$$

3. The function and the first three derivatives are
$f(x) = 2x^3 - 3x^2 + 1,\ f'(x) = 6x^2 - 6x, f''(x) = 12x - 6, f'''(x) = 12.$
Evaluated at $x = 0$:
$f(0) = 1,\ f'(0) = 0,\ f''(0) = -6,$ and $f'''(0) = 12$. The third-degree Maclaurin Polynomial is

$$1 + 0(x) - 6\left(\frac{x^2}{2!}\right) + 12\left(\frac{x^3}{3!}\right) = 1 - 3x^2 + 2x^3.$$

5. The function and the first three derivatives are

$f(x) = \ln(x + 1), f'(x) = \dfrac{1}{x + 1}, f''(x) = \dfrac{-1}{(x + 1)^2}$, and

$f'''(x) = \dfrac{2}{(x + 1)^3}$. Evaluated at $x = 0$:

$f(0) = 0,\ f'(0) = 1,\ f''(0) = -1,$ and $f'''(0) = 2$. The third-degree Maclaurin Polynomial is

$$0 + 1(x) - 1\left(\frac{x^2}{2!}\right) + 2\left(\frac{x^3}{3!}\right) = x - \frac{x^2}{2} + \frac{x^3}{3}.$$

7. The function and the first three derivatives are
$f(x) = (x+1)^{3/2}, f'(x) = \frac{3}{2}(x+1)^{1/2}, f''(x) = \frac{3}{4}(x+1)^{-1/2}$, and
$f'''(x) = \frac{-3}{8}(x+1)^{-3/2}$. Evaluated at $x = 0$: $f(0) = 1$, $f'(0) = \frac{3}{2}$,
$f''(0) = \frac{3}{4}$, and $f'''(0) = \frac{-3}{8}$. The third-degree Maclaurin Polynomial
is $1 + \frac{3}{2}(x) + \frac{3}{4}\left(\frac{x^2}{2!}\right) - \frac{3}{8}\left(\frac{x^3}{3!}\right) = 1 + \frac{3}{2}x + \frac{3}{8}x^2 - \frac{1}{16}x^3$.

9. The function and the first three derivatives are
$f(x) = \frac{1}{1-x}, f'(x) = \frac{1}{(1-x)^2}, f''(x) = \frac{2}{(1-x)^3}$, and
$f'''(x) = \frac{6}{(1-x)^4}$. Evaluated at $x = 0$: $f(0) = 1, f'(0) = 1, f''(0) = 2$,
and $f'''(0) = 6$. The third-degree Maclaurin Polynomial is

$$1 + 1(x) + 2\left(\frac{x^2}{2!}\right) + 6\left(\frac{x^3}{3!}\right) = 1 + x + x^2 + x^3.$$

11. The function and the first three derivatives are
$f(x) = \ln x$, $f'(x) = \frac{1}{x}$, $f''(x) = \frac{1}{x^2}$, and $f'''(x) = \frac{2}{x^3}$. Evaluated at
$x = 1$: $f(1) = 0, f'(1) = 1, f''(1) = -1$, and $f'''(1) = 2$.
The third-degree Taylor Polynomial is

$$0 + 1(x-1) - 1\left(\frac{(x-1)^2}{2!}\right) + 2\left(\frac{(x-1)^3}{3!}\right)$$

$$= (x-1) - \frac{1}{2}(x-1)^2 + \frac{1}{3}(x-1)^3.$$

13. The function and first four derivatives are
$f(x) = \frac{1}{3-x}$, $f'(x) = \frac{1}{(3-x)^2}$, $f''(x) = \frac{2}{(3-x)^3}$, $f'''(x) = \frac{6}{(3-x)^4}$
and $f^{iv}(x) = \frac{24}{(3-x)^5}$. Evaluated at $x = 2$:
$f(1) = 1, f'(1) = 1$, $f''(1) = 2$, $f'''(0) = 6$, and $f^{iv}(1) = 24$.
The fourth-degree Taylor Polynomial is

$$1 + 1(x-2) + 2\left(\frac{(x-2)^2}{2!}\right) + 6\left(\frac{(x-2)^3}{3!}\right) + 24\left(\frac{(x-2)^4}{4!}\right)$$

$$= 1 + (x-2) + (x-2)^2 + (x-2)^3 + (x-2)^4.$$

15. The function and the first six derivatives are $f(x) = e^x$, $f'(x) = e^x$, $f''(x) = e^x$, $f'''(x) = e^x$, $f^{iv}(x) = e^x$, $f^{v}(x) = e^x$, and $f^{vi}(x) = e^x$. Evaluated at $x = 1$: $f(1) = e$, $f'(1) = e$, $f''(1) = e$, $f'''(1) = e, f^{iv}(1) = e, f^{v}(1) = e$, and $f^{vi}(1) = e$.

The sixth-degree Taylor Polynomial is

$$e + e(x-1) + e\left(\frac{(x-1)^2}{2!}\right) + e\left(\frac{(x-1)^3}{3!}\right) + e\left(\frac{(x-1)^4}{4!}\right)$$

$$+ e\left(\frac{(x-1)^5}{5!}\right) + e\left(\frac{(x-1)^6}{6!}\right)$$

$$= e + e(x-1) + e\frac{(x-1)^2}{2} + e\frac{(x-1)^3}{6} + e\frac{(x-1)^4}{24}$$

$$+ e\frac{(x-1)^5}{120} + e\frac{(x-1)^6}{720}.$$

17. The function and the first four derivatives are $f(x) = e^{2x}, f'(x) = 2e^{2x}, f''(x) = 4e^{2x}, f'''(x) = 8e^{2x}$, and $f^{iv}(x) = 16e^{2x}$. Evaluated at $x = 3$: $f(3) = e^6, f'(3) = 2e^6$, $f''(3) = 4e^6, f'''(3) = 8e^6$, and $f^{iv}(3) = 16e^6$.

The fourth-degree Taylor Polynomial is

$$e^6 + 2e^6(x-3) + 4e^6\left(\frac{(x-3)^2}{2!}\right) + 8e^6\left(\frac{(x-3)^3}{3!}\right)$$

$$+ 16e^6\left(\frac{(x-3)^4}{4!}\right)$$

$$= e^6 + 2e^6(x-3) + 2e^6(x-3)^2 + \frac{4e^6(x-3)^3}{3} + \frac{2e^6(x-3)^4}{3}.$$

19. The function and the first three derivatives are $f(x) = x^{3/2}$, $f'(x) = \frac{3}{2}x^{1/2}$, $f''(x) = \frac{3}{4}x^{-1/2}$, and $f'''(x) = \frac{-3}{8}x^{-3/2}$. Evaluated at $x = 4$: $f(4) = 8, f'(4) = 3, f''(4) = \frac{3}{8}$, and $f'''(4) = \frac{-3}{64}$. The third-degree Taylor Polynomial is

$$8 + 3(x-4) + \frac{3}{8}\left(\frac{(x-4)^2}{2!}\right) - \frac{3}{64}\left(\frac{(x-4)^3}{3!}\right)$$

$$= 8 + 3(x-4) + \frac{3}{16}(x-4)^2 - \frac{1}{128}(x-4)^3.$$

21. Let $f(x) = x^{1/2}$, then the first few derivatives are
$f'(x) = \frac{1}{2}x^{-1/2}, f''(x) = \frac{-1}{4}x^{-3/2}$, and $f'''(x) = \frac{3}{8}x^{-5/2}$.
Evaluated at $x = 1$: $f(1) = 1, f'(1) = \frac{1}{2}, f''(1) = \frac{-1}{4}$, and $f'''(1) = \frac{3}{8}$.
The Polynomial takes the form

$$f(x) = 1 + \frac{1}{2}(x-1) - \frac{1}{8}(x-1)^2 + \frac{1}{16}(x-1)^3 + \cdots$$

One seeks $f(1.04)$ to three decimal places. Add terms until their difference is less than 0.0005. The estimate is accurate to three decimal places.

Therefore, $f(1.04) = 1 + \frac{1}{2}(0.04) - \frac{1}{8}(0.04)^2 \approx 1.020$.

23. Let $f(x) = x^{1/3}$, then the first few derivatives are $f'(x) = \frac{1}{3}x^{-2/3}$, and $f''(x) = \frac{-2}{9}x^{-5/3}$. The function and its derivatives at $x = 27$ yield $f(27) = 3, f'(27) = \frac{1}{27}, f''(27) = \frac{-2}{2187}$. The Taylor Polynomial takes the form

$$f(x) = 3 + \frac{1}{27}(x-27) - \frac{1}{2{,}187}(x-27)^2 + \cdots$$

One seeks $f(26.98)$ to three decimal places. Add terms until their difference is less than 0.0005. The estimate is accurate to three decimal places.

Therefore, $f(26.98) = 3 + \frac{1}{27}(-0.02) \approx 2.999$.

25. The function and the first three derivatives are $f(x) = e^x, f'(x) = e^x$, $f''(x) = e^x$, and $f'''(x) = e^x$. Evaluated at $x = 0$: are $f(0) = 1, f'(0) = 1$, $f''(0) = 1$, and $f'''(0) = 1$. The third-degree Taylor (Maclaurin) Polynomial is $f(x) = 1 + x + \frac{x^2}{2} + \frac{x^3}{6}$ and
$f(0.1) = 1 + (0.1) + \frac{(0.1)^2}{2} + \frac{(0.1)^3}{6} \approx 1.105$

EXERCISES 9.3

1. The function and its first few derivatives are $f(x) = \frac{1}{1+x}$, $f'(x) = -(1+x)^{-2}$, $f''(x) = 2(1+x)^{-3}$, $f'''(x) = -6(1+x)^{-4}$, and $f^{iv}(x) = 24(1+x)^{-5}$. Evaluated at $x = 0$, respectively: 1, −1, 2, −6,

and 24. Therefore, the expansion yields

$$1 - 1(x) + 2\left(\frac{x^2}{2!}\right) - 6\left(\frac{x^3}{3!}\right) + 24\left(\frac{x^4}{4!}\right) + \cdots n!\left(\frac{(-x)^n}{n!}\right) + \cdots$$

Simplifying yields

$$\frac{1}{1+x} = 1 - x + x^2 - x^3 + x^4 + \cdots + x^n + \cdots$$

$$= \sum_{n=0}^{\infty} (-x)^n \quad \text{for} \quad -1 < x < 1.$$

3. The function and its first few derivatives are $f(x) = \dfrac{1}{1+2x}$, $f'(x) = -2(1+2x)^{-2}$, $f''(x) = 8(1+2x)^{-3}$, $f'''(x) = -48(1+2x)^{-4}$, and $f^{iv}(x) = 384(1+2x)^{-5}$. Evaluated at $x = 0$, respectively: 1, −2, 8, −48, and 384. Therefore, the expansion yields

$$1 - 2(x) + 8\left(\frac{x^2}{2!}\right) - 48\left(\frac{x^3}{3!}\right) + 384\left(\frac{x^4}{4!}\right) + \cdots n!(-2)^n\left(\frac{x^n}{n!}\right) + \cdots$$

Simplifying yields

$$\frac{1}{1+2x} = 1 + (-2x) + (-2x)^2 + (-2x)^3 + (-2x)^4$$

$$+ \cdots + (-2x)^n + \cdots$$

$$= \sum_{n=0}^{\infty} (-2x)^n.$$

(Notice that it is simpler to replace x in Exercise 1 by $(2x)$, and the series is only valid for $\dfrac{-1}{2} < x < \dfrac{1}{2}$.)

5. The function and its first few derivatives are $f(x) = \dfrac{1}{1+x^2}$,

$f'(x) = -2x(1+x^2)^{-2}$, $f''(x) = \dfrac{6x^2-2}{(1+x^2)^3}$, $f'''(x) = \dfrac{24x-24x^3}{(1+x^2)^4}$, and $f^{iv}(x) = \dfrac{120x^4-240x^2+24}{(1+x^2)^5}$. Evaluated at $x = 0$, respectively: 0, −2, 0, and 24. Therefore, the expansion yields

$$1 + 0(x) - 2\left(\frac{x^2}{2!}\right) + 0\left(\frac{x^3}{3!}\right) + 24\left(\frac{x^4}{4!}\right) + \cdots n!\left(\frac{(-x^2)^n}{n!}\right) + \cdots$$

⟶

Simplifying yields

$$\frac{1}{1+x^2} = 1 + (-x^2) + (-x^2)^2 + \cdots + (-x^2)^n + \cdots = \sum_{n=0}^{\infty} (-x^2)^n.$$

(Notice that it is simpler to replace x by x^2 in Exercise 1 for $-1 < x < 1$.)

7. The function and its first few derivatives are $f(x) = \ln(1 + x)$, $f'(x) = \dfrac{1}{1+x}, f''(x) = -(1+x)^{-2}, f'''(x) = 2(1+x)^{-3}$, and $f^{iv}(x) = -6(1+x)^{-4}$. Evaluated at $x = 0$, respectively: 0, 1, -1, 2, and -6. Therefore, the expansion yields

$$0 + 1(x) - \left(\frac{x^2}{2!}\right) + 2\left(\frac{x^3}{3!}\right) - 6\left(\frac{x^4}{4!}\right) + \cdots (-1)^{n+1}(n-1)!\left(\frac{x^n}{n!}\right) + \cdots$$

Simplifying,

$$\ln(1+x) = x - \frac{x^2}{2} + \frac{x^3}{3} - \frac{x^4}{4} + \cdots + (-1)^{n+1}\frac{x^n}{n} + \cdots = \sum_{n=1}^{\infty} (-1)^{n+1}\frac{x^n}{n}$$

or $\displaystyle\sum_{n=0}^{\infty} (-1)^n \frac{x^{n+1}}{n+1}$

Since $\displaystyle\int \frac{1}{1+x}\,dx = \ln(1+x)$, Exercise 1 could be integrated term by term to yield the series.

9. The function and its first few derivatives are $f(x) = 4e^{x/2}, f'(x) = 2e^{x/2}, f''(x) = e^{x/2}, f'''(x) = \dfrac{1}{2}e^{x/2}$, and $f^{iv}(x) = \dfrac{1}{4}e^{x/2}$. Evaluated at $x = 0$, respectively: 4, 2, 1, $\dfrac{1}{2}$, and $\dfrac{1}{4}$. Therefore, the expansion yields

$$4 + 2(x) + \left(\frac{x^2}{2!}\right) + \frac{1}{2}\left(\frac{x^3}{3!}\right) + \frac{1}{4}\left(\frac{x^4}{4!}\right) + \cdots 4\left(\frac{1}{2}\right)^n\left(\frac{x^n}{n!}\right) + \cdots$$

Simplifying,

$$4e^{x/2} = 4 + 2x + \frac{x^2}{2} + \frac{x^3}{12} + \frac{x^4}{96} + \cdots + (4)\frac{\left(\frac{x}{2}\right)^n}{n!} + \cdots$$

$$= 4\sum_{n=0}^{\infty} \frac{(1/2)^n x^n}{n!}.$$

(Also in the series for e^x replace x by $x/2$ and multiply by 4.)

11. The function and its first few derivatives are $f(x) = xe^x - x$, $f'(x) = (x+1)e^x - 1, f''(x) = (x+2)e^x, f'''(x) = (x+3)e^x$, and $f^{iv}(x) = (x+4)e^x$. The function and these derivatives evaluated at $x = 0$ are 3, 0, 2, 3, and 4. Therefore, the expansion yields

$$0 + (0)(x) + 2\left(\frac{x^2}{2!}\right) + 3\left(\frac{x^3}{3!}\right) + 4\left(\frac{x^4}{4!}\right) + \cdots n\frac{(x)^n}{n!} + \cdots$$

$$= \sum_{n=2}^{\infty} \frac{x^n}{(n-1)!}.$$

13. Use the expansion for e^x so

$$\frac{1}{2}(e^x + e^{-x}) = \frac{1}{2}\left[1 + x + \frac{x^2}{2!} + \frac{x^3}{3!} + \cdots + \frac{x^n}{n!} + \cdots\right]$$

$$+ \frac{1}{2}\left[1 + (-x) + \frac{(-x)^2}{2!} + \frac{(-x)^3}{3!} + \cdots + \frac{(x)^n}{n!} + \cdots\right]$$

$$= \frac{1}{2}\left[2 + 0\,(x) + 2\left(\frac{x^2}{2!}\right) + 0\left(\frac{x^3}{3!}\right) + 2\left(\frac{x^4}{4!}\right) + \cdots\right]$$

$$= 1 + \frac{x^2}{2!} + \frac{x^4}{4!} + \cdots \frac{x^n}{n!} \quad n = 0, 2, 4 \ldots$$

$$\text{or } \frac{1}{2}(e^x + e^{-x}) = \sum_{n=0}^{\infty} \frac{(x)^{2n}}{(2n)!}$$

15. The function and its first four derivatives of $\ln x$ at $x = 1/2$ are $f\left(\frac{1}{2}\right) = \ln\left(\frac{1}{2}\right)$

$$f'(x) = \frac{1}{x}, \quad f'\left(\frac{1}{2}\right) = \frac{1}{2} \quad f''(x) = \frac{-1}{x^2}, \quad f''\left(\frac{1}{2}\right) = -4$$

$$f'''(x) = \frac{2}{x^3}, \quad f'''\left(\frac{1}{2}\right) = 16 \quad f^{iv}(x) = \frac{-6}{x^4}, \quad f^{iv}\left(\frac{1}{2}\right) = -96$$

Substituting yields,

$$\ln\left(\frac{1}{2}\right) + 2\left(x - \frac{1}{2}\right) - 4\frac{\left(x - \frac{1}{2}\right)^2}{2!} + 16\frac{\left(x - \frac{1}{2}\right)^3}{3!} - 96\frac{\left(x - \frac{1}{2}\right)^4}{4!} + \cdots$$

$$\ln\left(\frac{1}{2}\right) + \sum_{n=1}^{\infty} \frac{(-1)^{n+1}(2)^n\left(x - \frac{1}{2}\right)^n}{n}.$$

17. Exercise 9.3.1 yielded (for $-1 < x < 1$)

$$\frac{1}{1+x} = 1 - x + x^2 - x^3 + x^4 + \cdots + x^n + \cdots = \sum_{n=0}^{\infty} (-x)^n$$

$$\int \frac{1}{1+x} dx = \int (1 - x + x^2 - x^3 + x^4 + \cdots + (-x)^n + \cdots) dx$$

$$= \int \sum_{n=0}^{\infty} (-x)^n dx$$

$$\ln|x+1| + C = x - \frac{x^2}{2} + \frac{x^3}{3} - \frac{x^4}{4} + \frac{x^5}{5} - \cdots + \frac{(-1)^n x^{n+1}}{n+1} + \cdots$$

If $x = 0$ then $\ln 1 + C = 0$, so $C = 0$.

$$\ln|x+1| = \sum_{n=0}^{\infty} \frac{(-1)^n x^{n+1}}{n+1}$$

EXERCISES 9.4

1. The integral test yields

$$\int_3^{\infty} \frac{1}{x} dx = \lim_{t\to\infty} \int_3^t \frac{1}{x} dx = \lim_{t\to\infty} (\ln t - \ln 3) \to \infty.$$

The series is divergent.

3. The integral test yields $\int_1^{\infty} \frac{\ln x}{x} dx = \lim_{t\to\infty} \int_1^t \frac{\ln x}{x} dx.$

Using substitution to integrate yields $\lim_{t\to\infty} \left[\frac{(\ln t)^2}{2} - \frac{(\ln 1)^2}{2}\right] \to \infty.$

The series is divergent.

5. The integral test yields $\int_2^{\infty} \frac{1}{x(\ln x)^3} dx = \lim_{t\to\infty} \int_2^t \frac{1}{x(\ln x)^3} dx.$ Using substitution to integrate yields $\lim_{t\to\infty} \left[-\frac{1}{2(\ln t)^2} + \frac{1}{2(\ln 2)^2}\right] = \frac{1}{2(\ln 2)^2}.$

The limit is $\frac{1}{2(\ln 2)^2}$ so the series is convergent.

7. The integral test yields $\int_2^{\infty} 2xe^{-x^2} dx = \lim_{t\to\infty} \int_2^t 2xe^{-x^2} dx.$

Using substitution to integrate yields $\lim_{t\to\infty} [-e^{-t^2} + e^{-4}] = e^{-4}.$

The limit is e^{-4} so the series is convergent.

9. First, $n^3 + 1 > n^3$ and $\sqrt{n^3+1} > \sqrt{n^3}$. It follows that $\frac{1}{\sqrt{n^3+1}} < \frac{1}{\sqrt{n^3}}$. Therefore, $\sum_{n=1}^{\infty} \frac{1}{\sqrt{n^3+1}} < \sum_{n=1}^{\infty} \frac{1}{\sqrt{n^3}}$. The summation $\sum_{n=1}^{\infty} \frac{1}{\sqrt{n^3}} = \sum \frac{1}{n^{3/2}}$ is a p series that converges since $3/2 > 1$. A series which is term for term less than a convergent series is convergent. Therefore, $\sum_{n=1}^{\infty} \frac{1}{\sqrt{n^3+1}}$ is convergent.

11. First, $n^2 - 1 < n^2$ and $\sqrt{n^2-1} < \sqrt{n^2}$. It follows that $\frac{1}{\sqrt{n^2-1}} > \frac{1}{\sqrt{n^2}}$. Therefore, $\sum_{n=1}^{\infty} \frac{1}{\sqrt{n^2-1}} > \sum_{n=1}^{\infty} \frac{1}{\sqrt{n^2}}$.

The summation $\sum_{n=1}^{\infty} \frac{1}{\sqrt{n^2}} = \sum_{n=1}^{\infty} \frac{1}{n}$ is the divergent harmonic series ($p = 1$). Any series that is term for term greater than a divergent series is also divergent. Therefore, $\sum_{n=2}^{\infty} \frac{1}{\sqrt{n^2-1}}$ is divergent.

13. First, $n^3 + 1 > n^3$ and $\frac{1}{n^3+1} < \frac{1}{n^3}$. Multiplying by n yields $\frac{n}{n^3+1} < \frac{n}{n^3} = \frac{1}{n^2}$. Therefore,

$$\sum_{n=1}^{\infty} \frac{n}{n^3+1} < \sum_{n=1}^{\infty} \frac{n}{n^3} = \sum_{n=1}^{\infty} \frac{1}{n^2}$$ a convergent p series ($p = 2$).

Therefore, $\frac{1}{2} + \frac{2}{9} + \frac{3}{28} + \frac{4}{65} + \cdots + \frac{n}{n^3+1} + \cdots$ is convergent.

15. First, $\ln n < n^{1/2}$ and therefore, $\frac{\ln n}{n^2} < \frac{n^{1/2}}{n^2} = \frac{1}{n^{3/2}}$.

It follows that $\sum_{n=2}^{\infty} \frac{\ln n}{n^2} < \sum_{n=2}^{\infty} \frac{1}{n^{3/2}}$.
A convergent p series with $p = 3/2 > 1$.

A series term for term less than a convergent series also converges. Therefore, $\frac{\ln 2}{4} + \frac{\ln 3}{9} + \frac{\ln 4}{16} + \cdots + \frac{\ln n}{n^2} + \cdots$ is convergent.

17. Using an integral test to evaluate $\int_2^{\infty} \frac{1}{x(\ln x)^p} dx$. Using a substitution yields $\lim_{t\to\infty} \left(\frac{t}{(1-p)\, t^p} - \frac{\ln 2}{(1-p)(\ln 2)^p} \right)$.

⟶

When $p < 1$ the limit is infinite, so the integral is divergent and when $p > 1$, $\int_2^{\infty} \frac{1}{x(\ln x)^p}dx = \frac{1}{1-p}\frac{\ln 2}{(\ln 2)^p}$ and is convergent.

19. Here, $a_n = \frac{3^n}{n!}$ and $a_{n+1} = \frac{3^{n+1}}{(n+1)!}$. The ratio test relies on the limit:

$$\lim_{n\to\infty}\left|\frac{a_{n+1}}{a_n}\right| = \lim_{n\to\infty}\left|\frac{\frac{3^{n+1}}{(n+1)!}}{\frac{3^n}{n!}}\right| = \lim_{n\to\infty}\left|\frac{3^{n+1}}{(n+1)!}\cdot\frac{n!}{3^n}\right|$$

$$= \lim_{n\to\infty}\left|\frac{3}{n+1}\right| = 0 < 1.$$

The series $\sum_{n=0}^{\infty}\frac{3^n}{n!} = \frac{3}{1} + \frac{9}{2} + \frac{27}{6} + \cdots$ is convergent.

21. Here, $a_n = \frac{4^n}{n^3}$ and $a_{n+1} = \frac{4^{n+1}}{(n+1)^3}$. The ratio test relies on the limit:

$$\lim_{n\to\infty}\left|\frac{a_{n+1}}{a_n}\right| = \lim_{n\to\infty}\left|\frac{\frac{4^{n+1}}{(n+1)^3}}{\frac{4^n}{n^3}}\right| = \lim_{n\to\infty}\left|\frac{4^{n+1}}{(n+1)^3}\cdot\frac{n^3}{4^n}\right|$$

$$= \lim_{n\to\infty}\left|4\frac{n^3}{(n+1)^3}\right| = 4 > 1.$$

The series $\sum_{n=1}^{\infty}\frac{4^n}{n^3} = \frac{4}{1} + \frac{16}{8} + \frac{64}{27} + \cdots$ is divergent.

23. Here, $a_n = \frac{(-1)^n 6^n}{n!}$ and $a_{n+1} = \frac{(-1)^{n+1}6^{n+1}}{(n+1)!}$. The ratio test relies on the limit:

$$\lim_{n\to\infty}\left|\frac{a_{n+1}}{a_n}\right| = \lim_{n\to\infty}\left|\frac{\frac{(-1)^{n+1}6^{n+1}}{(n+1)!}}{\frac{(-1)^n 6^n}{n!}}\right| = \lim_{n\to\infty}\left|\frac{(-1)(6)n!}{(n+1)!}\right|$$

$$= \lim_{n\to\infty}\left|\frac{-6}{n+1}\right| = 0 < 1.$$

The series $\sum_{n=0}^{\infty}\frac{(-1)^n 6^n}{n!} = \frac{1}{1} - \frac{6}{1} + \frac{36}{2} + \cdots$ is convergent.

EXERCISES 9.5

1. The arithmetic sequence $1+4+7+\cdots+40$ has $n=14$, $a=1$, and $d=3$. Its sum is $14\left(1+\frac{13}{2}(3)\right)=287$.

3. The arithmetic sequence $11+15+19+\cdots+35$ has $n=7$, $a=11$, and $d=4$. Its sum is $7\left(1+\frac{6}{2}(4)\right)=161$.

5. The geometric sequence $\frac{3}{2}+\left(\frac{3}{2}\right)^2+\left(\frac{3}{2}\right)^3+\left(\frac{3}{2}\right)^4+\left(\frac{3}{2}\right)^5$ has

$a=3/2$, $r=3/2$, and $n=5$. Its sum is $\frac{3}{2}\left(\frac{1-\left(\frac{3}{2}\right)^5}{1-\frac{3}{2}}\right)=\frac{633}{32}$.

7. The geometric sequence $\frac{1}{5}+\frac{1}{5^2}+\frac{1}{5^3}+\frac{1}{5^4}+\cdots+\frac{1}{5^{12}}$ has $a=1/5$,

$r=1/5$, and $n=12$. Its sum is $\frac{1}{5}\left(\frac{1-\left(\frac{1}{5}\right)^5}{1-\frac{1}{5}}\right)=\frac{61,035,156}{244,140,625}$.

9. $\frac{5}{2}-\frac{5^2}{2^4}+\frac{5^3}{2^7}-\frac{5^4}{2^{10}}+\cdots-\frac{5^{12}}{2^{34}}$ has $a=5/2$, $r=-5/8$, and $n=12$.

Its sum is $\frac{5}{2}\left(\frac{1-\left(-\frac{5}{8}\right)^{12}}{1-\left(-\frac{5}{8}\right)}\right)\approx 1.533$.

11. $10+12+14.4+\cdots+24.8832$ has $a=10$, $r=1.2$, and $n=6$.

Its sum is $10\left(\frac{1-(1.2)^6}{1-(1.2)}\right)=\frac{62,062}{625}$.

13. Here, $a=3$, $r=5$, and $n=6$. Its sum is $3\left(\frac{1-(5)^6}{1-(5)}\right)=11,718$.

15. Here, $a=3$, $r=-2$, and $n=10$. Its sum is

$3\left(\frac{1-(-2)^{10}}{1-(-2)}\right)=-1023$.

17. Here, $a=1$ and $r=0.51$, so one seeks n such that

$1\left(\frac{1-(0.51)^n}{1-(0.51)}\right)<2$. Therefore,

⟶

$$2.0408163\,\lfloor 1-(0.51)^n \rfloor < 2$$
$$\lfloor 1-(0.51)^n \rfloor < 0.98$$
$$(0.51)^n > 0.02$$
$$n < \frac{\ln(0.02)}{\ln(0.51)} = 5.80984$$

After the fifth day, a dose is withheld to avoid an overdose.

SUPPLEMENTARY EXERCISES CHAPTER 9

1. Here, $a = \frac{2}{9}$ and $r = \frac{1}{3^3} = \frac{1}{27}$. Therefore, $S = \dfrac{\frac{2}{9}}{1-\frac{1}{27}} = \frac{3}{13}$.

3. Here, 0.363636 … is rewritten as $\frac{36}{10^2} + \frac{36}{10^4} + \frac{36}{10^6} + \cdots$

Therefore, $a = \frac{36}{100}$ and $r = \frac{1}{100}$ and $S = \dfrac{\frac{36}{100}}{1-\frac{1}{100}} = \frac{36}{99} = \frac{4}{11}$.

5. Here, $3.0\overline{45}$ is rewritten as $3 + \frac{45}{10^3} + \frac{45}{10^5} + \frac{45}{10^7} + \cdots$

Therefore, $a = \frac{45}{1000}$ and $r = \frac{1}{100}$ and

$$S = 3 + \frac{\frac{45}{1000}}{1-\frac{1}{100}} = 3 + \frac{5}{110} = \frac{335}{110} = \frac{67}{22}.$$

7. $\sum_{x=1}^{\infty}\left(\frac{3}{5}\right)^x$ Here, $a = \frac{3}{5}$ and $r = \frac{3}{5}$ and $S = \dfrac{\frac{3}{5}}{1-\frac{3}{5}} = \frac{3}{2}$.

9. The function and its first three derivatives are

$f(x) = \frac{2}{4-x}$, $f'(x) = 2(4-x)^{-2}$. The function and its first few derivatives are $f(x) = \frac{2}{4-x}$, $f'(x) = 2(4-x)^{-2}$, $f''(x) = 4(4-x)^{-3}$, and $f'''(x) = 12(4-x)^{-4}$. Evaluated at $x = 3$, respectively: 2, 2, 4, and 12. The expansion is

$$2 + 2(x-3) + 4\left(\frac{(x-3)^2}{2!}\right) + 12\left(\frac{(x-3)^3}{3!}\right)$$
$$= 2 + 2(x-3) + 2(x-3)^2 + 2(x-3)^3$$

11. The function and its first few derivatives are $f(x) = 2e^{x/5}$, $f'(x) = \frac{2}{5}e^{x/5}$, $f''(x) = \frac{2}{25}e^{x/5}$, $f'''(x) = \frac{2}{125}e^{x/5}$, and $f^n(x) = \frac{2}{5^n}e^{x/5}$. Evaluated at $x = 0$, respectively: 2, $\frac{2}{5}$, $\frac{2}{25}$, $\frac{2}{125}$ and continues as $\frac{2}{5^n}$. The expansion is

$$2 + \left(\frac{2}{5}\right)(x) + \left(\frac{2}{25}\right)\left(\frac{x^2}{2!}\right) + \left(\frac{2}{125}\right)\left(\frac{x^3}{3!}\right) + \cdots \left(\frac{2}{5^n}\right)\left(\frac{x^n}{n!}\right) + \cdots$$
$$= 2\sum_{n=0}^{\infty} \frac{(1/5)^n x^n}{n!}.$$

13. The integral test yields $\int_1^{\infty} \frac{x^2}{x^3+2}dx = \lim_{t\to\infty} \int_1^t \frac{x^2}{x^3+2}dx$. Using u substitution to integrate: $\lim_{t\to\infty}\left[\frac{1}{3}\ln|t^3+2| - \frac{1}{3}\ln 3\right]$. Since the limit approaches infinity, the series is divergent.

15. Here, $a_n = \frac{(n+1)(-1)^n}{n!}$ and $a_{n+1} = \frac{(n+2)(-1)^{n+1}}{(n+1)!}$. The ratio test relies on the limit:

$$\lim_{n\to\infty}\left|\frac{a_{n+1}}{a_n}\right| = \lim_{n\to\infty}\left|\frac{\frac{(n+2)(-1)^{n+1}}{(n+1)!}}{\frac{(n+1)(-1)^n}{n!}}\right|$$
$$= \lim_{n\to\infty}\left|\frac{(n+2)(-1)^{n+1}}{(n+1)!}\cdot\frac{n!}{(n+1)(-1)^n}\right|$$
$$= \lim_{n\to\infty}\left|\frac{(-1)(n+2)}{(n+1)^2}\right| = 0 < 1.$$

The series $\sum_{n=1}^{\infty}\frac{(n+1)(-1)^n}{n!}$ is convergent.

17. Here, $a = 5,\ d = 4$, and $n = 11$ so

$$5 + 9 + 13 + \cdots 45 = 11\left(5 + \frac{10}{2}(4)\right) = 275.$$

19. Here, $a = 5$, $r = 3$, and $n = 5$ yields $5\left(\dfrac{1-(3)^5}{1-3}\right) = 605.$

21. Here, $a = 9$, $r = 4$, and $n = 10$ yields $9\left(\dfrac{1-(4)^{10}}{1-4}\right) = 3{,}145{,}725.$

CHAPTER 10

APPLICATIONS TO PROBABILITY

EXERCISES 10.1

1. a) continuous c) discrete e) continuous
b) discrete d) continuous

3. a) $0.14 + 0.16 + 0.30 + 0.50 = 1.1$, which exceeds unity, so it is not a probability distribution.
b) $0.4 + 0.3 + 0.2 + 0.1 = 1$, each probability is nonnegative, so it is a probability distribution.
c) $0.12 + 0.18 + 0.14 + 0.16 + 0.20 + 0.25 + 0.05 = 1.1$, which exceeds unity, and is not a probability distribution.

5. a) $P(x \geq 53) = 0.05 + 0.30 + 0.10 + 0.20 = 0.65$
b) $P(x > 55) = 0.30 + 0.10 + 0.20 = 0.60$
c) $P(x \leq 58) = 0.20 + 0.15 + 0.05 + 0.30 + 0.10 = 0.80$
d) $P(52 \leq x < 60) = 0.15 + 0.05 + 0.30 + 0.10 = 0.60$
e) $P(x \leq 57) = 0.20 + 0.15 + 0.05 + 0.30 = 0.70$
f) $P(x = 59) = 0$

Solutions Manual to Accompany Fundamentals of Calculus, First Edition. Carla C. Morris and Robert M. Stark.

7. There are two properties to check: First, that $f(x)$ is nonnegative on the interval [1, 8], which it is. Next, that $\int_1^8 \frac{1}{7}\,dx = \frac{1}{7}x\Big|_1^8 = \frac{8}{7} - \frac{1}{7} = 1.$

9. There are two properties to check: First, that $f(x)$ is nonnegative on the interval [0, 10] which it is. Next, that

$$\int_0^{10} \frac{1}{50}x\,dx = \frac{x^2}{100}\Big|_0^{10} = \frac{100}{100} - \frac{0}{100} = 1.$$

11. There are two properties to check: First, that $f(x)$ is nonnegative on the interval [0, 1] which it is. Next, that $\int_0^1 3x^2dx = x^3\big|_0^1 = 1 - 0 = 1.$

13. There are two properties to check: First, that $f(x)$ is nonnegative on the interval [0, 1] which it is. Next, that $\int_0^1 4x^3dx = x^4\big|_0^1 = 1 - 0 = 1.$

15. There are two properties to check: First, that $f(x)$ is nonnegative on the interval $[0, \infty)$ which it is. Next, that

$$\int_0^{\infty} 3e^{-3x}dx = \lim_{t\to\infty}\int_0^t 3e^{-3x}dx = \lim_{t\to\infty}(-e^{-3t} + 1) = 1.$$

17. $\int_2^7 \frac{1}{7}\,dx = \frac{1}{7}x\Big|_2^7 = \frac{7}{7} - \frac{2}{7} = \frac{5}{7}.$

19. $\int_3^7 \frac{1}{50}x\,dx = \frac{1}{100}x^2\Big|_3^7 = \frac{49}{100} - \frac{9}{100} = \frac{40}{100} = \frac{2}{5}.$

21. $\int_{0.5}^1 3x^2dx = x^3\Big|_{0.5}^1 = 1 - \frac{1}{8} = \frac{7}{8}.$

23. $\int_{0.1}^{0.8} 4x^3dx = x^4\Big|_{0.1}^{0.8} = 0.4096 - 0.001 = 0.4095$

25. $\int_{1/3}^{5} 3e^{-3x}dx = -e^{-3x}\Big|_{1/3}^{5} = e^{-1} - e^{-15} \approx 0.3679$

EXERCISES 10.2

1. $E(x) = 50(0.20) + 100(0.10) + 150(0.30) + 200(0.40) = 145$

$$\sigma^2 = (50 - 145)^2(0.20) + (100 - 145)^2(0.10) + (150 - 145)^2(0.30) + (200 - 145)^2(0.40) = 3225$$

$$\sigma = \sqrt{3225} = 56.789$$

3. $E(x) = 1(0.15) + 4(0.15) + 7(0.25) + 10(0.20) + 12(0.25) = 7.5$

$$\sigma^2 = (1 - 7.5)^2(0.15) + (4 - 7.5)^2(0.15) + (7 - 7.5)^2(0.25) + (10 - 7.5)^2(0.20) + (12 - 7.5)^2(0.25) = 14.55$$

$$\sigma = \sqrt{14.55} = 3.814$$

5. The mean, $E(x)$; variance, σ^2, and standard deviation, σ are

$$E(x) = \int_1^8 \frac{1}{7}x\,dx = \left.\frac{x^2}{14}\right|_1^8 = \frac{64}{14} - \frac{1}{14} = \frac{63}{14} = 4.5.$$

$$\sigma^2 = \left(\int_1^8 \frac{1}{7}x^2\,dx\right) - (4.5)^2 = \left.\frac{x^3}{21}\right|_1^8 - (4.5)^2 = \frac{343}{84} = \frac{49}{12}, \text{ and}$$

$$\sigma = \sqrt{\frac{49}{12}} = 2.0207.$$

7. The mean, $E(x)$; variance, σ^2, and standard deviation, σ are

$$E(x) = \int_0^{10} \frac{1}{50}x^2\,dx = \left.\frac{x^3}{150}\right|_0^{10} = \frac{1000}{150} - \frac{0}{150} = \frac{20}{3}$$

$$\sigma^2 = \left(\int_0^{10} \frac{1}{50}x^3\,dx\right) - \left(\frac{20}{3}\right)^2 = \left.\frac{x^4}{200}\right|_0^{10} - \left(\frac{400}{9}\right) = \frac{50}{9}, \text{ and}$$

$$\sigma = \sqrt{\frac{50}{9}} = 2.357.$$

9. The mean, $E(x)$; variance, σ^2, and standard deviation, σ are

$$E(x) = \int_0^1 3x^3\,dx = \left.\frac{3x^4}{4}\right|_0^1 = \frac{3}{4} - 0 = \frac{3}{4}.$$

$$\sigma^2 = \left(\int_0^1 3x^4\,dx\right) - \left(\frac{3}{4}\right)^2 = \left.\frac{3x^5}{5}\right|_0^1 - \left(\frac{9}{16}\right) = \frac{3}{5} - \frac{9}{16} = \frac{3}{80}, \text{ and}$$

$$\sigma = \sqrt{\frac{3}{80}} = 0.1936.$$

11. The mean, $E(x)$; variance, σ^2, and standard deviation, σ are

$$E(x) = \int_0^1 4x^4 dx = \left.\frac{4x^5}{5}\right|_0^1 = \frac{4}{5}.$$

$$\sigma^2 = \left(\int_0^1 4x^5 dx\right) - \left(\frac{4}{5}\right)^2 = \frac{2x^6}{3}\Big|_0^1 - \left(\frac{16}{25}\right) = \frac{2}{75}, \text{ and}$$

$$\sigma = \sqrt{\frac{2}{75}} = 0.1633.$$

13. The mean, $E(x)$; variance, σ^2, and standard deviation, σ are

$$E(x) = \int_0^\infty 3xe^{-3x} dx = \lim_{t\to\infty} \int_0^t 3xe^{-3x} dx$$

$$= \lim_{t\to\infty} \left[\left(-xe^{-3x} - \frac{e^{-3x}}{3}\right)\Big|_0^t\right] = \frac{1}{3}.$$

$$\sigma^2 = \left(\int_0^\infty 3x^2e^{-3x} dx\right) - \left(\frac{1}{3}\right)^2 = \lim_{t\to\infty} \left(\int_0^t 3x^2e^{-3x} dx\right) - \left(\frac{1}{9}\right)$$

$$= \lim_{t\to\infty} \left[\left(-x^2e^{-3x} - \frac{2xe^{-3x}}{3} - \frac{2e^{-3x}}{9}\right)\Big|_0^t - \left(\frac{1}{9}\right)\right] = \frac{1}{9} \text{ and}$$

$$\sigma = \frac{1}{3}.$$

EXERCISES 10.3

1. A bell curve sketch is useful. In this case, the probabilities are obtained directly from a Standard Normal Table.

a) $P(0 \le z \le 1.47) = 0.4292$
b) $P(0 \le z \le 0.97) = 0.3340$
c) $P(-2.36 < z < 0) = 0.4909$
d) $P(-1.24 < z < 0) = 0.3925$
e) $P(-2.13 \le z \le 0) = 0.4834$
f) $P(-0.19 \le z \le 0) = 0.0753$

3. a) This includes the entire upper half of the distribution. Add 0.5000 and 0.4394 ($P(-1.55 < Z < 0)$) to yield 0.9394.

b) This is in the upper tail, so $0.5000 - 0.4686 = 0.0314$.

c) This is the lower tail, so $0.5000 - 0.4292 = 0.0708$.

d) This is the entire lower half of the curve plus $P(0 < Z < 1.30)$. Therefore, $0.5000 + 0.4032 = 0.9032$.

5. a) $P(460 \leq x \leq 640) = P\left(\frac{460-550}{100} \leq x \leq \frac{640-550}{100}\right)$

$$= P(-0.90 \leq z \leq 0.90)$$

$$= 0.3159 + 0.3159 = 0.6318.$$

b) $P(x \leq 730) = P\left(z \leq \frac{730-550}{100}\right)$

$$= P(z \leq 1.80) = 0.5000 + 0.4641 = 0.9641.$$

c) $P(x \geq 410) = P\left(z \geq \frac{410-550}{100}\right)$

$$= P(z \geq -1.40) = 0.5000 + 0.4192 = 0.9192.$$

7. Here, $\mu = 128.4$.
One seeks $P(x \leq 128) \leq 0.01$. The z score corresponding to this probability is -2.33.
Therefore,

$$-2.33 = \frac{128 - 128.4}{\sigma}$$

Solving yields a value of 0.171674 for the standard deviation σ.

SUPPLEMENTARY EXERCISES CHAPTER 10

1. a) $0.13 + 0.17 + 0.35 + 0.45 = 1.1$, which exceeds unity; it is not a probability distribution.

b) $0.33 + 0.27 + 0.22 + 0.18 = 1$, and each probability is nonnegative, so it is a probability distribution.

3. a) $P(15)+P(16)+P(18)+P(20)=0.30+0.05+0.20+0.10=0.65$

b) $P(18) + P(20) = 0.20 + 0.10 = 0.30$

c) $P(12) + P(15) = 0.20 + 0.30 = 0.50$

d) $P(x \leq 18) = 1 - P(20) = 1 - 0.10 = 0.90$

5. First, $f(x)$ is nonnegative on [0, 1]. Second,

$$\int_0^1 5x^4\,dx = x^5\big|_0^1 = 1 - 0 = 1.$$

Therefore, $f(x)$ satisfies both criteria for a probability density function.

7. The mean, $E(x)$; variance, σ^2, and standard deviation, σ are

$$E(x) = \int_0^1 5x^5 dx = \left.\frac{5x^6}{6}\right|_0^1 = \frac{5}{6}.$$

$$\sigma^2 = \int_0^1 5x^6 dx - \left(\frac{5}{6}\right)^2 = \left.\frac{5x^7}{7}\right|_0^1 - \left(\frac{5}{6}\right)^2 = \frac{5}{7} - \frac{25}{36} = \frac{5}{252} \text{ and}$$

$$\sigma = \sqrt{\frac{5}{252}} = 0.141.$$

Lightning Source UK Ltd.
Milton Keynes UK
UKOW06f0015180216

268620UK00001B/94/P